Guiding Emotions

Emotions play a crucial role in the design process, yet junior design students often struggle to manage their emotional changes while creating. This title bridges the gap in design education by illuminating how emotional intelligence serves as a powerful tool for inspiration and decision-making. This investigation takes a new perspective on emotions in design projects, contrasting it with traditional approaches that prioritise technical skills and reasoning over emotional awareness. The focus of this title is to deliver accessible strategies to monitor and manage the reader's emotional responses during projects.

This book presents a Phase-by-Phase Emotional Configuration Model alongside a taxonomy of tactics for concept creation, selection and prototyping. Through empirical research, 40 essential emotions linked to design were identified and validated using advanced emotion tracking techniques, including mobile applications and facial expression analysis. Case studies and reflective activities provide practical guidelines for emotional self-regulation throughout the design journey. By integrating these tools into their practice, readers will develop the ability to harness emotions as creative catalysts rather than obstacles, ultimately producing more resonant and effective design solutions.

Guiding Emotions: An Emotion-Tracking Approach to Rectify Design Experience and Process is for professionals, educators, students and researchers of design, human factors and creative industries.

Design, Emotion and Creativity
Series Editor: Amic G. Ho

This book series explores the close relationships between design, emotional experiences and the creative process. The books presented here comprehensively understand how these three aspects intertwine and influence one another while offering theories, insights and strategies for designers, researchers, and creative students alike. Themes within this series include the role of emotion in design, investigations into the creative process, innovative design methods and techniques, design thinking, user experience (UX), computing and emotion recognition and the future trends and challenges for emotion in the design process. Aimed at those who aim to design more creatively and innovatively, please get in touch to discuss your book idea for this series.

Emotion in the Design Process
Intrinsic Factors on Emotion Management for Decision-Making
Amic G Ho

Guiding Emotions
An Emotion-tracking Approach to Rectify Design Experience and Process
Amic G Ho

Guiding Emotions

An Emotion-Tracking Approach to Rectify Design Experience and Process

Amic G. Ho

CRC Press
Taylor & Francis Group
Boca Raton London New York

CRC Press is an imprint of the
Taylor & Francis Group, an **informa** business

Designed cover image: Amic G. Ho

First edition published 2027
by CRC Press
2385 NW Executive Center Drive, Suite 320, Boca Raton FL 33431

and by CRC Press
4 Park Square, Milton Park, Abingdon, Oxon, OX14 4RN

CRC Press is an imprint of Taylor & Francis Group, LLC

ISBN: 9781041255864 (hbk)
ISBN: 9781041209393 (pbk)
ISBN: 9781003747345 (ebk)

DOI: 10.1201/9781003747345

Typeset in Times
by Deanta Global Publishing Services, Chennai, India

This work was supported by the Faculty Development Scheme
(Grant number UGC/FDS16/E06/19)
the University Grants Committee, Hong Kong SAR

Contents

About the Author

Amic G. Ho is a design scholar and researcher specialising in the intersection of emotion, art and design. His work focuses on the role of emotion in the creative process, aiming to create holistic experiences that integrate art, design, education and technology. Dr Ho holds a PhD in Design Studies from the University of Hong Kong, along with an MEd and MDes, and has published extensively in academic books, journals and conference proceedings. He has led collaborative projects funded by local and international grants, producing impactful findings. As a book series editor at Taylor & Francis Group, Dr Ho has overseen influential works on Design, Emotion and Creativity. His recent publications include *Emotion in the Design Process: Intrinsic Factors on Emotion Management for Decision-making* (2023) and *Emotion Guideline Workbook: Managing the Design Process* (2024) published by Taylor & Francis Group. Dr Ho also serves as Chair and Editor for the International Conference on Human Factors in Communication of Design (AHFE) and the International Symposium on Creative Technology and Digital Media (AHFE). Through his research, publications and leadership roles, Dr Ho continues to advance the understanding of emotion in design, fostering interdisciplinary collaboration and innovation.

Preface

The purpose of this book is rooted in longstanding curiosity about how students entering the junior ranks of the design profession develop the ability to manage the complexities associated with the design process, while developing the capacity to recognise and appreciate the subjective experience of those for whom they design. Ultimately, the primary goal of this work is to identify design education strategies that foster the development of affective skills (i.e., empathy and emotional awareness), so that designers can respond more thoughtfully and creatively to the needs of others.

Design transcends mere problem-solving; it is a human-centred and complex pursuit that requires synthesising varied knowledge, including psychological and physiological aspects. Novice designers often struggle to connect with users whose experiences differ greatly from their own, such as older adults or individuals with disabilities. This difficulty arises not from a lack of interest but from inadequate tools to recognise and utilise their emotions throughout the design process.

The body of research represented by this book evolved out of a pedagogical question: Can systematic, research-based methods assist students in incorporating emotion into their design processes in a manner that improves both their learning and their design outcomes? The research presented in this book drew upon the literature related to emotion and cognition as well as previous research related to simulation-based design and emotion tracking technologies to provide insight into how emotions occur during the design process, how they influence decision-making, and how they may be directed versus suppressed. The focus of the research was on junior design students as they represent a critical developmental stage at which students' habits, attitudes and self-conceptualisations will evolve to become ingrained in their professional practice.

To address the aforementioned research questions, the research employed a multi-stage methodology that integrated design assignments, lecture content, scaffolded activities, mobile emotion tracking technology, survey instruments, group discussions, lab logs and interviews. In two major studies, students completed design assignments while systematically documenting and contemplating their emotional states, and then applied principles for integrating emotion into their design process. Utilising this approach allowed for both qualitative and quantitative assessments of how specific emotional variables interact with phases such as research, ideation, material selection and prototype development.

The results indicate that emotions are not tangential to the design process but are instead intricately intertwined with cognitive processing and decision-making across all phases of the design process. Additionally, certain emotional states, when acknowledged and regulated, appeared to enhance the effectiveness of decision-making and facilitate more empathetic responsive design engagements with end-users. While historically, emotions such as confusion and frustration have typically been viewed as impediments to successful design process completion, this research explores the conditions under which these emotions can be leveraged as positive indicators of the design process.

This book is targeted towards three distinct audiences: junior design students seeking to better understand and control their emotions in the context of design; design educators seeking practical approaches to embedding emotional awareness and empathy into design curricula; and design researchers and practitioners interested in the emerging area of design and emotion. For students, the proposed guidelines provide concrete questions and frameworks to aid them in recognising, adjusting and utilising their emotional responses as part of their design practice. For educators and researchers, the work provides a theoretical and methodological framework for future investigation into affective competencies in design education.

Ultimately, this work aspires to contribute to an optimised design process wherein self-awareness, empathy and emotional intelligence are recognised as fundamental aspects of professional competence, rather than optional personal characteristics. Through emphasising emotion as both an object of study and a pedagogical resource, this book seeks to inspire a generation of designers who are not only technically competent and creative, but also more attuned to the diverse emotional experiences of the people they design for, and more reflective about their own.

Acknowledgements

Grateful acknowledgement is made to the University Grant Committee for providing me with a research grant to conduct this study.

The work described in this paper was fully supported by a grant from the Research Grants Council of the Hong Kong Special Administrative Region, China [UGC/FDS16/E06/19].

***

Special thanks to Ms Ruth Chau P. W. the research assistant in this project, who devoted unlimited time and effort to supporting completion of this research over weekdays, weekends and holidays.

*

I would like to thank my PhD supervisor, Professor Michael Siu K. W. for providing me with the opportunity to explore knowledge in the field of design and emotion, which helped me to gain additional knowledge to further develop this aspect into design education.

*

Finally, I would like to sincerely thank the Design and Emotion Society and all its founding members for highlighting this field of research and contributing valuable insights to the field.

1 Introduction

1.1 OVERVIEW

Design is recognised as a complex, creative and open-ended skill that involves the integration of many different types of knowledge (Van Dooren et al., 2014). In the case of designing for older populations, the designer must concurrently consider not only typical issues such as space, form, image, light, colour, composition and material but must also balance knowledge about age-related conditions and the sociology of ageing within the context of the specific project. Despite this complexity, expert designers experience the design process as an automatic process involving unconscious steps or actions based on common practice or routine (Van Dooren et al., 2014). For novice designers, this process is anything but common or routine. Students must be trained to juggle many domains of information and recognise relations between all the aspects involved (Eastman, 2001; Lawson, 2006; Van Dooren et al., 2014).

Given that empathy enables designers to achieve the above goals, that undergraduate programmes exist to train entry-level design practitioners and that emotions also govern the extent to which learners attend to germane learning content, it seems critical to explore how to teach affective skills such as empathy at the undergraduate design level. Moreover, given its heavy focus on human-centred design, interior design education presents itself as an ideal context in which to study the impact of pedagogical methods on empathetic skills development. Empathy is a core affective skill for designers, who must be able to understand the users for whom they design (Esser 2018). Kouprie and Visser (2009) defined 'empathy' as the ability to feel 'as if' by taking on the viewpoint of another to relate to situations. Designers use empathy to identify and incorporate the thoughts, emotions and expectations of others to create effective design solutions (Altay et al., 2016).

Although some researchers (e.g., Lazarus, 1984; Schachter & Singer, 1962; Zajonc, 1998) have posited a strong connection between emotional experiences and the cognitive appraisal of the experiences, the precise role of specific emotions during the learning process remains under debate (Park et al., 2015). For example, some researchers indicated that negative and positive moods induced by emotional experiences can impact students' cognitive processing by impacting their ability to regulate abstract thinking and problem-solving strategy selection (Bless et al., 2006). Other researchers (e.g., D'Mello et al., 2014) pointed to the counterintuitive positive effect of negative emotions such as confusion. More work is needed to explore the impact of emotions on how design students learn to design productively and respond to emotional experiences, as well as the aspects of learning interventions that induce effective emotional experiences.

DOI: 10.1201/9781003747345-1

1.2 PROPOSED OBJECTIVES

The objectives of this study are as follows:

- To understand how emotions influence students' design processes with measurable tools.
- To investigate the emotional elements that may affect junior design students' abilities to manipulate their design process.
- To propose principles to guide junior design students in adjusting their emotional changes in the design process.
- To examine the effectiveness of principles aimed to guide junior design students in controlling their emotional changes in the design process.

1.3 DEFINITION OF THE RESEARCH PROBLEM

A common obstacle to many young design students' ability to create designs for users who experience the world differently than they do (i.e., because of some form of disability), particularly if these users represent the majority of a population, is developing a solid understanding of the physical, physiological and emotional experiences of their intended end-users, and going beyond what is simply observable to include the lived and affective aspects of daily life. Developing these capabilities will allow them to use design as a means of making decisions to help assure the safety and well-being of their end-users.

Professionals in design-related fields have demonstrated that simulation-based design methods are effective at combining cognitive processes with explicit knowledge products to increase and deepen designers' understanding of their users and users' needs; thus, there is a compelling educational question for educators of design students: can design educators adopt simulation design methods to enhance both cognitive and affective learning by increasing design students' self-awareness and empathy? The objective of this research project was to determine if design educators can utilise structured simulation design guidelines to enhance future designers' sensitivity to the variability in human emotional regulation abilities/limitations, and to assess how simulation-based design strategies may contribute to the development of design students' affective competencies.

1.4 RESEARCH QUESTIONS AND HYPOTHESES

This study explored the potential methods for junior design students with which to manage and introduce emotions into their design processes with accompanying guidelines to said methods. Both the positive and negative effects of emotional adjustments were explored to propose guidelines for junior design students. The type and function of the emotional elements that influence the internal and external factors of design processes were explored. Junior design students can apply the developed guidelines to assist them in managing their emotions as part of their design processes, thereby practically modifying their design process.

Research Questions

1. What emotions do Junior Design Students experience when they manipulate their design processes?
2. Junior Design Student's cognitive reflection of their design experiences as a way to identify, how their emotional states shift as they go through the design process demonstrates that emotions have an effect on the way their design experiences develop?
3. How is emotion a part of Junior Design Students' experience during the design process; and how does emotion affect their cognitive thinking throughout the design process?
4. How will the 'Principles for Design & Emotion Concept Configuration in the Design Process' help Junior Design Students?

1.5 SIGNIFICANCE OF THE STUDY

1.5.1 Impact and Objectives

This study is about assisting students who are in the early years of a design programme in dealing with both their emotional reactions and their design process. To help achieve this goal the researchers have four objectives. First, they want to find methods to include and integrate emotion into the design process so as to improve the decisions that students make during the design process, and ultimately improve the designs produced. Second, they intend to look at what types of emotional variables affect the way students perform and manage their design activities, and what role those variables play. Third, they plan to evaluate the best ways to manage student's emotions through all phases of the design process based on prior research related to the effects of various emotional aspects on students' decision-making processes at different points in time. Fourth, they hope to create practical strategies to assist junior design students in better understanding and dealing with their own emotional responses while engaged in the design process. The results from this research will contribute valuable knowledge to the field of design education, which will aid other educators in teaching students how to effectively manage their emotional responses while designing.

1.5.2 Educational Implications

1.5.2.1 Long-Term Impact

Concerning long-term impact, this research is significant in terms of introducing the 'design and emotion' concept into the study of the design process (Naqvi et al., 2006). This contribution to the existing knowledge will be provided to junior creative arts/design students, design educators, design researchers and design participants.

Applicable Guidelines for Junior Design Students

This research provides insights to help junior design students consider methods to modify their design processes through emotional management as part of their design process. While the appropriate emotions that are required in each stage of the design process are suggested, several guiding questions for managing their own emotional responses are proposed for those leading junior design students. The decision-making and performance of junior design students, as part of the design process, can be improved in this way. The self-regulated principles may improve design outcomes and students' academic achievements.

Introduce Emotional Concern in Design Education

In contrast to past research, which primarily focused on theoretical development, this study explored practical approaches for junior design students to introduce emotional responses into their design processes. Consequently, this study provides design educators with principles to help guide these students to adjust their emotional perspectives during the design process as part of their learning. The approaches for introducing emotional concerns into the design curriculum will be demonstrated.

New Insightful Development in Design Education Research

For design academia, this research provides a theoretical background and analysis to understand how particular emotional elements can be effective in the design process. The research results will be published to serve as guidelines for introducing emotion into the design process. These contributions will serve as a foundation for further developments and new insights in design and emotion studies.

1.5.3 Theoretical Implications

1.5.3.1 Optimised Design Process for the Design Industry

This study presents empirical research on methods to introduce emotional concerns into the design process of junior design students (who represent a new generation of designers) to achieve improved design outcomes. Different stages of the design process were reviewed in a step-by-step manner, for example, idea generation, material allocation and prototype generation. The particular ideal emotions were matched and the way they function in terms of enhancing the decision-making of these junior design students was established. The proposed guidelines provide design professionals with examples of the methods that can help them to improve how they manage their emotions. In addition, these guidelines can help professional design participants adjust their emotional responses. As such, the results of this study provide insights for design practitioners in terms of considering the concepts of design and emotion in their professional training.

1.6 THE RESEARCH PROCESS

1.6.1 Research Process

The entire research process was divided into two studies (Research Studies 1 and 2), the details of which are described in this section. It includes a design process, a lecture for introducing the concept of design and emotion, a focus group, an application of an emotion-tracking mobile application, an interview and the application of the CAPTIV L7000-Premier function (i.e., an emotion-tracking device set) to collect multi-feature data.

1.6.2 Research Study 1

1.6.2.1 Phase 1

Investigation of the emotional elements in the design process: First design project using an emotion-tracking mobile application

In the first stage of this research, all participants were invited to manipulate design projects individually with the use of an emotion-tracking application during the research, material allocation and idea generation phases. The objective of this stage was to discover junior design students' ideal emotional responses that enhance their judgement and evaluation as part of their design processes.

In Research Study 1, a mobile application, eMoods Bipolar Mood Tracker, was applied to obtain a systematic review to help recognise participants' featured emotional changes during their design practices. The mobile application helped participants record their opportunities and challenges related to design practices. The effectiveness of this mobile application has been proven by scholars who adopted this technology to observe mood changes among their participants (Caldeira et al., 2017). These scholars proposed that the eMoods Bipolar Mood Tracker is a practicable and appreciated method for tracking emotional changes through reporting, graphing and logging. Accordingly, this mobile application was used to ensure this research is comprehensive and to identify junior design students' ideal emotional responses that enhance their decision-making in their design processes.

1.6.2.2 Phase 2

Introduction of the proposed principles of controlling emotions in design: Lecture (scaffolding) for junior design students on accommodating emotions in the design process

In the second stage, the participants attended a lecture at which the findings derived in Phase 1 of Research Study 1 were shared. They obtained knowledge about the ideal emotional responses that enhance their decision-making in their design processes. They were subsequently invited to take part in several learning activities (scaffolding) to gain an understanding of some of the concepts of design and emotion as well as methods to introduce those concepts into their design processes. The objectives of this stage were to provide the participants with stimulation in the lecture, increased knowledge about design and emotion, and an improved understanding of the way they can recognise their emotions.

1.6.2.3 Phase 3

Questionnaire

In the third stage, the participants of each group were asked to review their design process using a questionnaire designed by the researcher to collect quantitative feedback. The objectives of this stage included understanding participants' recognition of their emotions and comparing the effectiveness of the general design process with a design process that introduces the proposed principles.

1.6.3 Research Study 2

1.6.3.1 Phase 1

Examination of the effectiveness of individual proposed principles: Second design project using an emotion-tracking mobile application

In this phase, another design project was assigned. Participants were invited to record their emotions with an emotion-tracking app during research, discussions and idea generation. The objective of this stage was to record the process of emotional changes visually and to examine the effectiveness of proposed principles individually.

1.6.3.2 Phase 2

Focus groups

In this stage, participants were invited to attend a 45-minute interview in 14 groups to deepen their understanding of their design process and experience under the influence of emotions and with the application of the emotion-tracking mobile application, CAPTIV L7000-Premier. During the focus group, participants were asked to review their experience of applying the proposed principles of emotion and design to their design process. The record in the emotion-tracking mobile application was reviewed. Additionally, to effectively understand participants' emotional adjustments in the focus group discussion, a system of recognising emotional responses, CAPTIV L7000-Premier, was applied. This application can track a wide variety of concerns, such as emotional responses, eye movement and self-control behaviours. The CAPTIV L7000-Premier application is functionally equivalent to selected emotional response tracking systems and was adopted in several existing studies that investigated the needs of children and adults.

Studies have highlighted the emotional response tracking technique as an essential technology for supporting the mental analysis algorithms of participants' cognitive systems. The technology enables the recording of accurate facial expressions and body language within the concept of emotional literacy (Noffke & Somekh, 2009). The effectiveness of the CAPTIV L7000-Premier application provided scientific data with which to examine the efficacy of the proposed principles individually in different stages of the design process.

1.6.3.3 Phase 3

Individual interviews and the reflective reviews in Lablog

In this stage, the three highest and three lowest achievers among the participants were invited to attend a 20-minute interview and a lablog review to better understand their emotional changes and the influence of emotions on their design process. This phase was audio-recorded and transcribed for coding by the researcher. The coding was analysed, both qualitatively and quantitatively, and coding schemes were developed.

Reflective review in participants' lablog

The lablog included a full record of the learning process and served as:

- a timeslot;
- a space in which to try new things;
- a reflective journal; and
- a thinking tool.

Each participant's lablog was a detailed account of the order and duration of all the aspects they focused on during the learning process, enabling them to use their time more effectively and, following reflection, could help to ensure effective time management in future projects. As a space in which to try new things, the lablog provides a notebook to record dates and places, people encountered, observations of phenomena, new thoughts and feelings, experiences and knowledge and journeys of discovery. As a reflective journal (Murty & Purcell, 2007), it serves as an invaluable database for the revision and improvement of students' thinking and problem-solving processes. It is a record of thoughts, impressions, opinions, action strategies, perceived outcomes and results. As a thinking tool, it provides a platform for the generation of ideas and is a portable, convenient platform to record mind maps, brainstorming lists, answers to questions raised, concept development and design analysis by revisiting and re-examining the creative process.

Among those aspects, lablog provides the best support for conditional knowledge, as it records all the information about the learning process. This can help students to know the reasons for making decisions and approaches to designing to apply declaratively and to gain procedural knowledge. This allows students to make decisions about resource allocation strategically; in turn, this will make their strategies or design solutions more effective. The thinking record ('lablog') of Leonardo da Vinci is regarded as a typical example of how a lablog can help learners process conditional knowledge. The adoption of the lablog in this research would eventually encourage participants to evaluate how the proposed principles could assist them in conducting the design process using metacognition. The concept of metacognition supported participants in providing insightful feedback within the field of design research. The process of planning, applying and evaluating (related to all thinking activities and problem-solving and evaluation procedures) using the lablog helped participants to enhance their knowledge and awareness and to regulate their learning progress through evaluation. From the preparation of the presentation on metacognition, insights into leading metacognitive strategies in learning were obtained. The lablog is the ultimate tool for tracking participants' emotional changes and finding answers to all possible questions related to this.

The knowledge about understanding design processes obtained by this research can be used for different tasks. Participants were able to recognise effective approaches for communicating with their peers in the group, who had similarly improved their interpersonal skills. This learning experience introduced the application of metacognitive control and self-regulatory processes to the participants' design learning processes. They applied cognitive processes to control and regulate their emotions during learning, including planning, monitoring and evaluating. The knowledge and experience obtained from this learning activity (improved understanding of their emotions and user experiences) included knowledge about and reflection on the investigation methods used to obtain the most important content, that is, deep knowledge versus partial knowledge, which supports improved design outcomes. The objectives of this stage were to discover the influence of emotional changes on participants' design processes and their knowledge about emotions in design. In addition, the effectiveness of the proposed principles for applying the knowledge of design and emotion collaboratively was examined, and the elements involved in the principles that motivate junior design students to learn were investigated.

2 Literature Review

2.1 OVERVIEW

Some scholars have expressed concerns about the relationship between humanity and design in recent decades. Some studies linked the concepts of emotion and design (Overbeeke & Hekkert, 1999) to understand how designs can provide a pleasant design experience (Pampliega & Marroquin, 1998) to users (Desmet, 2008; Herrmann & Schmidt, 2002). These investigations were subsequently developed into fundamental concepts of design and emotion, and the bulk of these studies were conducted according to user perspectives (Jordan, 2000; Lawson, 2006; Ortony et al., 1988). Concurrently, other studies underscored that emotional changes are human responses to the environment and influence people's judgement. Researchers have taken these insights into design studies as a reference and proposed that emotional adjustments may influence not only the experience of design (Lewis, 1995) but also the design process. They proposed that the influence of emotional modifications on judgement is affected by experience and knowledge (Ho & Siu, 2012). Consequently, junior designers/students experience a relatively more obvious influence as a result of emotional shifts during their design process than experienced design professionals. Some junior design students may encounter emotional changes during this manipulation (Pahl & Beitz, 1996; Brocato, 2009) via the process of innovation, and these emotional changes may influence their judgement and evaluation (Ho & Siu, 2012). This could indicate that the influence of emotional changes on decision-making is based on initiative. In addition, it indicates that emotional changes could influence the design process. However, few investigations observed changes in junior design students' emotions during the design process (Jensen, 1999) or how their emotional changes affect the design process (White et al., 2023).

Innovative design processes have been developed to facilitate a manageable design outcome. These design processes focused on, for example, management and materials allocation. However, junior design students still experienced difficulties manipulating the design process, even when they had learned management and materials allocation skills. Moreover, junior design students showed themselves to be weak in processing obtained information related to human interactions (Achmad et al., 2023). Their creative generation process involved not only rational but also logical considerations, as well as concerns related to their techniques, skills and personal experience. Moreover, the responses designers had towards the external environment or stimuli were regarded as emotional responses. Some scholars referred to these responses as 'intrinsic factors' (Ho & Siu, 2012). The emotional changes or responses related to these factors within affective systems influence designers' judgement and evaluation during the design process.

DOI: 10.1201/9781003747345-2

To become familiar with and understand how junior design students' emotions may affect their design process, their emotional changes (McClenaghan et al., 2007) were recorded and measured (Ho & Siu, 2012). Selected measurements adopted in the context of design and emotion, such as facial expressions (Desmet, 2015) and verbal feedback, provided helpful suggestions for recognising the emotional changes of designers. Based on a study conducted by Ho and Siu (2012) (a PhD study conducted by the author), theories related to the links between 'emotion', 'decision-making' and the 'design process' were developed. The author reviewed existing studies on 'design and emotion' and clarified theories pertaining to 'emotionalised design', 'emotional design' (Norman, 2004) and 'emotion design'. These theories linked relationships between designers, design outcomes and the user/audience within the 'design and emotion' context. These theories were then described using the 3E model framework. To further explain how emotions work alongside factors of the 'design process', another framework, the E-wheel model, was used to explain the interactions between internal factors, external factors and emotions and designers. The E-wheel model reflected that after recognising emotional changes, junior design students' emotions can be categorised. The impact of emotions on junior design students' performance in their design process can then be practically investigated. It was assumed that if junior design students adjust their emotional changes using appropriate methods, their decision-making abilities (Bailey et al., 2020). could be enhanced. Accordingly, junior design students can manipulate their design processes effectively, thereby optimising design outcomes. Thus, investigating practical guidelines to assist junior design students will help them become more aware of their own emotional changes. As such, introducing them to the design process in a manner that highlights optimised design outcomes is paramount.

This study aimed to explore the methods available to junior design students to manage and introduce emotions into their design processes using a set of relevant guidelines. The positive and negative effects of these emotional changes were explored to propose relevant guidelines. The type and function of the emotional elements that influence the internal and external factors of design processes were explored. Junior design students can adopt the developed guidelines to assist them in managing their emotions and practically modifying their design process.

2.2 DESIGN PROCESS

It is essential to understand the development and manipulation of the design process in design studies. There was a lack of systematic structure in the craftsmanship-emphasised period. Craftsmen reviewed their routine practices and gradually developed standard process stages, which subsequently served as a foundation for the principles of design outcomes. The objectives of design outcome approaches were based on two general principles: the division and specialisation of labour and the use of tools and equipment to create standardised design outcomes (Ho & Siu, 2012). Following the establishment of the Staatliches Bauhaus, which combined individual aesthetics with functionality, the principles of design outcomes were unified with artistic vision. In other words, the practical design outcome principles were revealed in institutional art education. Educators strove to create routine design outcome practices through several standard design phases. Hence, the principles of mass design outcomes provided the main elements for shaping the design process in design practice.

Lawson (2006) highlighted the importance of understanding the approach to governing knowledge about end-users and improving the design process. It was considered one of the most essential skills of the designer. Additional research was conducted to help design students develop their time management and project organisation abilities. Similarly, Cross and Sivaloganathan (2004) emphasised that the best approach to governing marketing risk is to enhance an understanding of and the interaction with end-users. They compared several formal methods, for example, marketing research and relatively direct interactions with end-users, setting up routine contact between designers and end-users and inviting innovative users to the marketing research team.

Studies have shown how the order of incorporating design requirement solutions can impact design outcomes. Jundanian (2007) investigated the fundamental design process concepts of 'design and emotion' according to user perspectives and noted that research in this regard was carried out using an experiment in which two junior design students were invited to design rooms for a residential area. They first drew sketches and floor plans to recognise potential problems related to the assigned residential area. Next, they were asked to improve their plans according to additional design requirements. These requirements were provided in different orders, and one of the participants had to redesign their existing plans. The completed design development sketches and final solutions were collected and analysed after the design projects were completed. Based on the results, Jundanian established hierarchies of design criteria, including primary structural design phases (e.g., pre-design, schematic design, design development, construction drawings, design outcome and construction administration). Although some of the categories are loosely defined, they are generally employed in design projects in real-world situations. These results show that design decisions are considered and re-considered repeatedly in different phases of the design process. A design decision may require a change in subsequent decisions and the creation of a new design solution. This information shows how the design process can be affected by the design requirements' sequence and the need for a unique approach to manipulate them effectively.

Several types of research have been conducted to help design students learn the basic principles of the design process as well as project management within the design industry (Murty and Purcell, 2007). Best (2006) noted that the iterative and non-linear character of design processes makes standardising them challenging. This is additionally influenced by the diverse demands of users, viewers and audiences. Best then questioned the function of design managers in the design process and how they might control the creative design process to foster inventive outcomes. Since it captures all the details of the learning process, the lablog was best positioned as a tool to support conditional knowledge. Students will benefit from knowing the methods (and when to use them) of applying information declaratively and procedurally. In this way, students can be encouraged to allocate resources strategically, helping them to successfully execute plans and creative ideas. A typical example of how a lablog can aid students in processing conditional information is Leonardo da Vinci's thinking journal. By implementing a lablog, participants will be encouraged to assess how the suggested principles may help them manage the design process using metacognition. In the realm of design research, the notion of metacognition can assist study participants in providing meaningful feedback. Thus, participants can improve their knowledge, awareness and ability to control their learning progress via assessment using the planning, application and assessment (for all thinking activities and problem-solving and evaluation procedures) processes in the lablog. Insights into the most effective metacognitive learning techniques can be gained via the preparation of a presentation on metacognition. Accordingly, the lablog is the best instrument with which to monitor participants' emotional changes and discover all pertinent information about them.

Naqvi et al. (2006) reviewed various design processes and summarised them alongside existing studies (Norman, 2004). In addition, user experience research considered that user feedback on design outcomes and conceptualisations indicates that users interpret their expectations (Norman, 2004). The researcher noted that this point of view was not only supported by the qualitative feedback from design experience questionnaires but also explained by the physical mesolimbic dopamine system in the human body. This system, hypothesised by the somatic-marker hypothesis, is a leading aspect of unconscious action bias and has increasingly become a topic of discussion in terms of playing a role in reward processing and decision-making. Participants were invited to present their responses to stimuli. Their responses were recorded and compared with predicted primary responses (Ho & Siu, 2012). The participants' feedback reflected that their responses shifted over time, from the receipt of primary reinforcement to being reinforced at the onset of the predictive cues. Using sophisticated computational techniques to create a model, activity within the mesolimbic dopamine system was found to signal reflected biased behaviour when comparing actual and predicted rewards.

Suchman (2003) investigated knowledge transformation in the design process and suggested that knowledge emerges within a design setting, while recognition arises from the design process in the same setting by applying risk management theory. After understanding the rationale for the design process and gaining knowledge about their target users and audience, businesses have adopted risk management as a supplementary tool to optimise the design process. In a real commercial environment, the most essential consideration of design outcome development is risk management because risk profiling (a tool to analyse all the potential threats that may be discussed by end-users and designers in advance) differs between projects, and potential profit is closely linked to risk management. Hence, Cross and Sivaloganathan (2004) emphasised that risk management should be implemented with systematic approaches to manage uncertainty related to potential threats. The researchers additionally stated that risk assessment and risk management are essential parts of the design process model. In the study, designers were expected to provide solutions to unique design problems. The methodologies employed by the designers were collected and summarised to achieve a comprehensive design aim.

The researchers proposed that design methods and methodologies were essential to establishing a design process model. Other critical aspects of the design process model included governance of the design process, consensus and a reduction in the design outcome time. According to Aken (2005), several models of the design process can be applied in the same period. However, their impact on the field of process design remains modest. Designers primarily rely on two types of design information when making decisions; one type is created using experience and knowledge based on evidence, whereas the other is established via rigorous and methodical investigation. Designers must be motivated and in charge of their own actions in addition to knowing the details of the design process. Aken (2005) questioned whether designers' personal experiences would affect the real design process.

Best (2006) claimed that users appraise design-oriented work through the emotional message that it conveys. The researcher developed a methodology to understand the relationship between the emotional responses among end-users and designers to a design, using an emotional interface as a demonstration. The study's hypothesis aimed to show whether users reflect emotion towards an interface through which they interact with colours, images and content. In addition, the study explored whether a natural change in focus from usability to emotion in the interface improves user performance. Best proposed that the design process acts as a tracking record to remind the design team to take action and to highlight the significance of their contributions. The design process was categorised into six parts: (1) defining the problem, (2) understanding the problem, (3) conceptualising the design, (4) developing the concept, (5) creating and evaluating a design prototype and (6) production and manufacturing. Tzvetanova (2007) investigated the design process phases in detail. In the 'defining the problem' phase, designers review the design brief to understand the needs of the end-users and to familiarise themselves with the research situation. In the 'understanding the problem' phase, the brief and the agreement on the responsibilities of each shareholder are finalised. In the 'conceptualising the design' phase, designers engage in creating the design, including reviewing its conceptualisation and agreeing on the concepts that are fit for development. In the 'developing the concept' phase, prototypes are developed and expanded. In the 'creating and evaluating the design prototype' phase, designers conduct the design in detail and engage in prototype evaluation, design outcome planning, design refinement, final presentation and design freeze and create a pre-design outcome prototype. In the 'producing and manufacturing' phase, the time required for creating tools, manufacture and project review is finalised.

Van Gorp and Adams (2012) investigated the fundamental design process concepts of 'design and emotion' according to user perspectives. The research was carried out using an experiment in which two junior design students were invited to design rooms for a residential area (Van Gorp & Adams, 2012). They first drew sketches and floor plans to identify the potential problems of the assigned residential area. Next, they were asked to improve their plans according to additional design requirements, which were provided in different orders and led to one participant having to redesign their existing plans. Completed sets of design development sketches and final solutions were collected and analysed once the design projects were completed. According to the research findings, Jundanian established more comprehensive hierarchies of design criteria, including primary structural design phases such as pre-design, schematic design, design development, construction drawings design outcome and construction administration. Although some of these categories are loosely defined, they are generally employed in design projects in real-world situations. This result indicates that design decisions are considered and re-considered repeatedly at different phases of the design process. One design decision can cause a change in subsequent decisions, and a new design solution may be required. This information shows how the design process can be affected by the design requirements' sequence and that unique techniques are needed to manipulate them effectively. Furthermore, some scholars investigated if supportive tools could ease the manipulation of the design process (Van Gorp & Adams, 2012).

2.2.1 The Design Process in Design Management

Supplementary management tools to improve the design process were investigated. Brissaud et al. (2003) explored whether a model/research methodology exists that could be further developed with obvious justification to support the design process. The researchers found that the method used by designers relies primarily on their observations and analyses. Accordingly, they proposed notetaking in the form of a reflective journal as a potentially effective tool. Sketches and materials recorded through notetaking were produced by the participating designers to suggest, explain, discuss and examine their perspectives on the constantly evolving design process. The records were validated by experimentation using methods applied in a laboratory setting and focused on the context of decision-making. This process involved a database that stored all the interactions and modules to process the application design in a design solution. Finally, an experiment to test the model was carried out. This involved a one-off design project for which designers were asked to create a mechanical system within a fixed time. The design process was divided into two successive joint meetings lasting three hours with breaks to conduct reviews. The experiment started with a list of end-user and audience needs and ended with sketches that would allow the design outcome to be manufactured.

Park (2007) further developed the relationship between systematic research and the design process. They stated that designers face challenges in improving the design process in terms of establishing effective communication. Designers are expected to lead interactive experiences among end-users, as well as create the design outcomes. Using advanced technology, the design process optimises the design experience to meet the needs of users. Chen and Chen (2004) analysed and extracted the key elements of the standard design process and design outcome management system (including feedback from users and the markets). To provide an executable and balanced process that is functional and managed capabilities, the two scholars adopted the concrete concepts of process evaluation. The operating procedure for the development of the adaptive design process into a holistic method was reviewed. Chen (2014) additionally investigated adaptive design process theory, which allows designers to easily create novel designs in the procurement of design patent protection by focusing on problem definition in the design process in the following years. To define the problem clearly, the scholars collected feedback from users. This approach can help design teams identify trends and the needs of their end-users, as well as the users' ability, and to understand the message or use of the design before the problem definition stage. The literature review and the case studies served as the foundation for the development of the design model. Feedback from expert designers was synthesised in several iterations. Group discussions were structured according to six topics and synthesised. Three disciplines were selected as demonstrations: engineering, mechanical engineering and architecture (Chen, 2014).

2.2.1.1 Designer Considerations Involving Emotions

To investigate the system for recording emotional changes, it is essential to construct an understanding of the design–emotion connection. Once an understanding of concepts concerning emotion (including both emotional and behavioural traits) is in place, Desmet (2015) suggested that design outcomes evoke emotions related to five categories, that is, utilitarian, aesthetic, social, surprise and curiosity. These emotions were shown to be effective in describing complex and often personal design outcomes. The five emotions reveal the links between thinking and information interpretation (i.e., design concept is developed as self-regulation). The 'design and emotion' concept assists designers in designing to evoke an emotional response. Desmet reviewed previous research and found that user–emotion links extend to commerce, services and marketing. To help designers design in a way that elicits emotion, specific study methods can be used that integrate users' emotional associations and relevant tools (Salminen et al., 2021). The design and emotion studies were categorised into user-centred, designer-centred, research-centred and theory-centred groups. The user-centred studies emphasise users' emotions and experiences as a driving force for innovation (Rasouli et al., 2022). These are closely associated with the social tactics an individual applies in their relationships with users (i.e., design concept is developed as reflecting empathy). Theory-based methods focus on research that generates insights from users/consumers to improve designs. They are closely linked with interpersonal communication that relies on verbal or non-verbal language (i.e., the concept is developed as reflecting design and emotional intelligence abilities). The research-centred method investigates designers as authoritative and conveying views on design experience. These aspects are closely linked with communication frequency, quality and impact and are all strongly reliant on the designers themselves (i.e., the concept is developed as reflecting self-awareness). This categorisation of the design and emotion studies reveals the mature and complex development of design and emotion theories.

The research-centre approach provided the foundation for Alaniz and Biazzo's (2019) investigation of the way emotion influences the creative process and emotion-centric design. It was challenging to create design outcomes with relatively subtle emotional qualities that are not dependent on the appeals and functions of design outcomes (Alaniz & Biazzo, 2019). Alaniz and Biazzo proposed that product designers and manufacturers may need to learn how to understand and express emotions through their products to address this issue successfully. Their study describes the invention of an emotion-driven innovation method to help product design teams visualise new product concepts. The research team employed Platts' (Alaniz & Biazzo, 2019) research approach, which includes four key components, namely a state-of-the-art study, process design, development and validation. An experimental survey was conducted among the global designer community to assess three key ideas (emotional framework, emotional occupations and human–product emotional interactions). The design community was polled using a crowdsourcing platform. Participants were instructed to upload a photo of a product they had designed or any product that they felt had been designed successfully and briefly explain it. Current studies, however, mainly focus on the appeal and function of design outcomes. Few studies on design and emotion have examined how emotion influences the creative process, design experience and interactions between designers and users/consumers (Brom et al., 2018).

2.2.2 The Design Process in Design Education

Some scholars in design education researched the application of the design process in design education to help junior designer students adopt systematic practices from the design industry. Cross and Sivaloganathan (2004) reviewed supplementary tools that improved the design process (risk management theory). The researchers followed Suchman's (2003) suggested approach, which proposes that recognition arises from the design process within a design situation. After gaining an understanding of the rationale for the design process and knowledge about target users and consumers, some businesses adopted risk management as a supplementary tool to optimise the design process. In a real commercial environment, an essential consideration of design outcome development is risk management. Risk profiling (a tool used to analyse potential threats that emerge from discussions between end-users and designers in advance) of projects and potential profits are closely linked to risk management. Cross and Sivaloganathan (2004) emphasised that risk management should be conducted using systematic approaches to manage uncertainty related to potential threats. Cross and Sivaloganathan additionally stated that risk assessment and risk management formulation are essential aspects of the design process model. In the study, designers were expected to provide solutions to unique design problems. The methodologies employed by the designers were collected and summarised to achieve a comprehensive design aim. As a result, the researchers proposed that design methods and methodologies are essential to consider when formulating a design process model. Other critical aspects of the design process model include governance of the design process, consensus and reducing the design production time.

In addition to the aforementioned design teaching and learning methodologies, several educators examined the development of new design education trends (Kaygan et al., 2020). Lim et al. (2021) proposed that before problem-solving, junior designers must comprehend the nature of the issue at hand. This is considered self-awareness and is a necessary skill for designers to have a successful start in their design work. Concerning junior designers knowing the nature of an issue, Brewer and Devnew (2022) shared the viewpoint that this is important and noted that social characteristics are no longer solutions to a design issue but rather strategic communication instruments. Bixler (2016) observed the shift of authority in teaching and learning from the instructor to the student. With this shift, planning, monitoring and evaluating instructional efficacy in the teaching and learning process places a significant emphasis on student motivation. Lage et al. (2022) showed that self-awareness and self-control are conflicting management skills for junior designers. However, junior designers often do not successfully manage their design processes. According to De Garrido, (2021), junior designers must address and react to the issue of designing an environment that supports, enhances and celebrates human activities. This idea results from a convergence of social, cultural and economic requirements that make it necessary for junior designers to take on many responsibilities and then make judgements based on these duties.

Denton et al. (2004) studied techniques used to convey the role of emotion in design to junior designers. The 'emotional domain' included a range of mental states. Both internal and external forces were shown to influence them. The effect was found to be decidedly individual, indicating that people may experience different emotions in reaction to the same stimuli. The research team found that the junior designers were not encouraged to communicate their emotions through project work; instead, they addressed them with instructors and among themselves. The emotional concerns that may have affected the design process were therefore disregarded. In addition, the researchers found that the junior designers did not seem to understand the full relevance and potential of emotion in design work. Endres et al. (2020) explored how emotional design may simultaneously promote learning. The researchers' cognitive-engagement hypothesis predicted that a change in emotional design would increase positive affect and engagement, thereby enhancing performance.

Ertmer and Russel (1995) noted that in the past, the bulk of creative art education had been aimed at giving designers clear instructions. There was a gap between the scenario in the classroom and the situation in the practice sessions. Students at creative art schools can process their work because they are required to achieve particular learning goals that have been set by their professors. Content and technique are two important components of the case method teaching approach. While lecturers are in charge of managing the subject of a project, they must also be acutely aware of how they deliver information. Teacher–student and student–student interactions can give rise to novel ideas about producing that can be conveyed, examined and ultimately altered in the design process.

2.3 DESIGN AND EMOTION CONCEPT CONSOLIDATION

The ability to understand user experience is a core skill for designers who must understand the target users of their design (Esser, 2018). Kouprie and Visser (2009) proposed that the ability to understand user experience, a concept described as 'empathy', is the capacity to feel 'as if one is the user' by taking on the user's viewpoint to relate to situations (Caldeira et al., 2017). The researchers emphasised that this ability is accumulated through an understanding of user experience and design practices. Experienced designers use their ability to understand user experience to identify and incorporate the thoughts, emotions and expectations of others to create effective design solutions (Altay et al., 2016).

The concepts of empathy and user experience are fundamental aspects of human-centred design but are not fully considered in all designs. Among the design domains, how the user's experience works out through the functions of the design outcomes is a concern. A designer's code of practice has been developed that notes how the profession is founded on human behaviour theories and describes the ability to create environments that address well-being. Its definitions reveal that the emotional needs of people are a fundamental responsibility of the designer (Council for Interior Design Accreditation, 2018; Roseman et al., 2021). Given that the ability to understand user experience enables designers to achieve their design goals, undergraduate programmes are essential training for junior design students. Emotions are one of the aspects governing the extent to which learners address germane learning content. Thus, it is critical to explore how to introduce affective skills such as empathy at the undergraduate design level. Moreover, given its heavy focus on human-centred design, design education presents itself as an ideal context in which to study the impact of pedagogical methods on empathetic skill development.

Scherer (2005) viewed theories about emotion broadly and proposed that emotion works as the interface between an organism and its environment. It constantly mediates between the individual and their environment, as the environment changes situations, events and, in most cases, the individual's behavioural responses. His concept broadened the direction of research on the way emotions relate to thought processes and decision-making.

In addition to Scherer's (2005) theory, Crilly et al. (2004) presented the concern that the appearance of design outcomes influences consumers' expectations of their emotional engagement with those design outcomes. Remaining cognisant of the aesthetic, semantic and symbolic aspects of the response to design outcomes and conceptualising them as part of the presented framework will help designers understand user responses to the design outcome's appearance. Reymen et al. (2006) investigated the appeal of design outcomes based on the viewpoint of state–transition theories. They modelled design outcomes according to the study participants' understanding of their personal characteristics, intuition and perception, creative thinking and other skills. This way, the design quality and outcomes could be improved (Dael et al., 2012). Design outcome appearance played a remarkable role in resolving the user responses, which included unarticulated user requirements and suggested expected design outcome qualities that were difficult to ascertain. As a result, the approach made it possible to assess whether the glamorous appearance of the design outcome was the sole factor or whether the function and marketing positioning of the design outcome played a role in creating a valued user experience (Chen et al., 2018).

There were difficulties in terms of standardising approaches to eliciting emotions in different cases (Brom et al., 2018; Reinholt et al., 2022). Emotional changes will be reflected in behaviour, physiological responses (Rendina, et al., 2019), emotions, aesthetic execution, gestures and verbal and physical presentation, as well as in specialised metrics such as eye-tracking and clickstream data. Beyond the appearance or functions of the design outcomes, a personal attachment aspect (Christner et al.,

2020) is also involved. Objects present in daily living are more than simply material possessions. Norman (2004) pointed out that humans take pride; this is not only to reflect their wealth or status but also to enjoy pleasant memories and the expressions that doing so elicits.

Based on cross-discipline studies on marketing, design and social science, Norman (2004) concluded that a design sets up the consumer's expression. This direction extended designers' concern for emotions, moods and feelings evoked by design outcomes. Tzvetanova (2007) explored the impact of the emotional response of customers and designers on the design. An approach was presented that could be used to investigate the emotional responses to a design on the part of users and designers. Based on the results, the approach with which the industry evaluates the impact of emotional responses between a customer and a designer to a design is understood. However, it is not comprehensive enough to cover the impact of the emotional response of the end-users and the designer on the design process.

Some design scholars obtained insights from psychological studies, such as the activation theory, in which several roles of emotions are applied to design aspects. Han and Park (2009) tried to use systematic research to clarify problem definitions during the first steps of the design process. Conversely, other companies in the industry tried to improve the design process model by studying how emotions impact other stages in the design process, such as concept development, detailed design and evaluation. The following concepts allowed us to learn how the industry developed a design process model.

Since emotional factors are crucial to design outcomes, knowledge of emotions and emotional factors in the context of design outcomes is important for designers to realise. Violante et al. (2019) introduced the background of emotion, emotional design theories and emotional design thinking. Using these aspects, designers could find a method of emotion classification, extract emotional factors and apply emotional theories in their design project. Based on the experience and achievement of forerunners in the emotional design field, a more systematic presentation and explanation of emotional design processes are possibly needed.

In a similar period, some designers found it essential to develop design outcomes that focused on the needs of the end-users. However, design outcomes and services have been developed with a focus on the needs of a specific user, thus increasing the risk of errors and misunderstandings (Marin-Alvarez et al., 2020). Among different design approaches, ergonomics was discussed alongside the needs of users. Ergonomics has a fundamental role. When the design of information is not addressed in a way appropriate for the cognitive needs of all users, errors and loss of users' time occur, as the documents do not invite reading. Instructions that are not understood can generate frustration, ambiguous information and excessive fatigue in those who interact with the devices. The study of ergonomics is essential; it is necessary to understand the users and study their needs, as there are limitations to their ability to understand what type of object is needed and all the characteristics and specifications it requires. Each user is different, and it is necessary to group them according to their personalities, limitations and needs.

The literature shows that it is essential to develop design outcomes that are focused on the needs of end-users. Few design outcomes and services have, however, been developed with a focus on the needs of a specific user, causing an increase in the risk of errors and misunderstandings. Some designers focus on improving a design by gaining a thorough understanding of the different users who interact with the design outcome. This includes information that only primary users (such as a doctor or health officer) may understand. However, other users (such as the patient and/or the patient's family) are not considered, which can lead to them not being able to understand the information that is presented to them. It is important to propose a taxonomy of the different users that interact with the design outcome and correspond to the results of the clinical analysis to develop design outcomes that are focused on their cognitive requirements.

Boess and Jong (2007) revealed that knowledge about target design users helps to set a clear design context, define problems and optimise design processes. They studied how designers used information about design outcomes and whether they were able to select the most useful information to create design outcomes that met users' needs. This process helped them to design empathically and to review how they evaluated their designs during the design process. From previous studies, the scholars learned that people expect design outcomes and services to work. They used Microsoft's package for the Windows Vista operating system in their demonstration. They employed emotional mapping to enlist both in-person and online respondents. Participants were recruited according to independent variables that included geographic location, type of buyer and gender. Respondents rated different designs according to various attributes, ranking them along a scale. The results were then statistically analysed and tracked.

2.3.1 Design and Emotion Concept

To investigate the emotional influence on the management of the design process, it is essential to understand the concept of emotions. Scherer and Ekman (2014) pointed out that the definition of emotion is broad. They developed a sequence to describe emotions in a relatively systematic manner. For example, arousal levels were employed to describe an individual's way of interacting through emotional responses. An individual always communicates with the environment's components as conditions and occurrences change. This interaction influences the person's behavioural responses. In other words, emotional responses are transmittable and influence human decisions and reactions. Applying the influence of emotion in the design process, Bongiovanni and Louis (2021)) suggested a design technique based on state–transition theories. Lesser-known skills, such as intuition, perception and creative thinking were used to simulate the design process. Bongiovanni and Louis proposed user feedback techniques for improving a product's quality and design process. User feedback suggested comprehensive design output characteristics and redefinable design requirements. The research team additionally pointed out that appearance influenced the users' emotions. The dazzling look of the design output was not the sole component in producing a valued user experience; utility and commercial placement also contributed. Thus, design outputs influence users' emotional connection with goods. Considering the aesthetic, semantic and symbolic elements of the users' reactions can help designers understand how customers react to design outputs.

2.3.2 Development of the Design and Emotion Concept

The Design and Emotion Society was founded in 1999. In the same year, major papers predicted the subsequent extensive interest in emotion. In *The Experience Economy*, Pine and Gilmore (1999) predicted the rise of an experience-based economy and the collapse of goods and services. The authors stressed that organisations must create engaging experiences to succeed. In *The Dream Society*, Jensen (1999) anticipated the commercialisation of emotions. Schmitt (1999) introduced experiential marketing to emphasise sensory responses and emotions instead of features and benefits. Schmitt stressed the relevance of emotions in product development, consumer engagement and business relationships.

Several visionary design articles showed a developing grasp of user emotions. Oughton (2022) convincingly argued that poorly designed user interfaces of most technological equipment (e.g., videocassette recorders, automobile alarms, software programmes) make users feel inept and unhappy. Picard (2003) suggested in *Affective Computing* that emotional competencies benefit computerised systems. In *Hertzian Tales*, Dunne (1999) suggested that junior designers of electronic devices should broaden their aesthetic considerations. He advised using industrial design to improve our relationship with artificial technology and stressed that industrial designs may improve our lives more than current designs. Gaver et al. (1999) helped junior designers understand the experiences of hard-to-reach user groups and encouraged them to apply this knowledge. Jordan (2000) stressed the need for enjoyable products, noting that previous usability-based design methodologies had limited products and devalued people by only considering their physical and cognitive abilities. He proposed a pleasure-based approach to human factors that comprehensively examines the connection between people and commodities and assesses design excellence based on the larger links between products and their users.

These studies suggest a growing interest in the emotional aspects of making, buying and consuming products. However, these aspects have not been embraced in education and practice. Junior designers have little information to consider the experiential impact of their designs. Some junior designers may try to understand the significance of emotions, but many view them as intuitive and elusive to analyse or predict. It can be difficult for junior designers to manipulate the design process; they may lack experience dealing with emotional changes and intuitive aspects throughout the design process, in addition to having to learn techniques to control the design variables using logical reasoning. Additionally, if they have no emotional concerns, they may be unable to recognise their own emotional changes. Thus, they may find it difficult to practise design and need instruction to improve their motivation, assessment and decision-making skills. Instead of evaluating design skill and craftsmanship, it is essential to create tools, methodologies and insights to help junior designers manipulate their design process more effectively by understanding and resolving the design processes' emotional consequences.

2.3.3 Emotion-Related Experience in Design Studies

In addition to concerns about user needs, Reymen et al. (2006) suggested an alternative design process based on the viewpoint of state–transition theories. The researchers modelled the design process according to the understanding between the study's teammates of their personal characteristics, intuition and perception, creative thinking and other skills. This helped to improve the quality of designs and the design process. Design outcome appearance played an important role in resolving user responses. The user responses provided unarticulated user requirements and suggested expected design outcome qualities that were difficult to ascertain. Judgements included that the glamorous appearance of design outcomes is not the sole factor in creating a positive user experience but that the functions and marketing positioning of the design outcomes also play a role. The appearance of design outcomes was shown to influence consumer expectations concerning their emotional engagement with the products. Remaining cognisant of the aesthetic, semantic and symbolic aspects of their responses, and conceptualising these as part of the presented framework, can help designers gain a deeper understanding of users' responses to design outcomes' appearance.

Wang and Zhou (2020) investigated emotional design and new design outcome ideas. As consumer demand has become increasingly personalised and diverse, consumption levels and aesthetic concepts are constantly being improved. The development of manufacturing and information industries has allowed people to experience material results while paying much attention to their emotional needs. To conduct a thorough investigation of the design outcome development paradigm from psychological and emotional perspectives (Despeisse, 2018), this research presents a decision-making approach for design outcome design that is driven by the user's Kansei image (the term 'Kansei' comes from Japanese aesthetics and can mean 'sensibility' or 'empathy'. In visual arts and design, a Kansei picture is an artistic production that uses visual media to create and embody emotions, moods or aesthetic sensitivities in the observer) via the application in the Electronic Gaming Machines (EGM) and the International Gaming Awards (IGA). Specific to this study is an analysis of the reasons for what users consider attractive factors in design outcomes, the upper perceptual image and the specific characteristics of design outcomes. These were observed using the expert interview method developed by EGM to create a database of user images and the specific characteristics of design outcomes. Based on the fuzzy Kano model, to obtain accurate user satisfaction in terms of perceivable design outcome needs, the Kansei image attributes of users were analysed. Observable image factors with high user satisfaction were extracted and used as the carrier of user interaction evaluation in IGA, thereby narrowing the scope of design outcome by extensively evaluating and modelling the components required by users in the evaluation process. The adaptive value of evolutionism was then automatically calculated by the system after being weighted in accordance with the value of the user assessment.

This strategy not only lessened the strain placed on users throughout the design outcome development process but also significantly increased the accuracy with which users evaluated new design outcomes. Furthermore, throughout the development of the manufacturing process, consumers' preferred items were transformed into ideas that were investigated.

2.4 PSYCHOLOGICAL STUDIES OF EMOTION

Interplay exists between emotion, memory, rational thinking, decision-making and behaviour. From a contemporary scientific perspective, emotion is considered an individual's cognitive representation of their physiological state (Gerald, 2010). Thinking and feeling are, on a physiological level, aspects of a single process occurring in the nervous system, itself a palimpsest of ancestral elements that are highly parallel and reflexively linked (Salas et al., 2012). All expressions will to some extent provide information based on social context. Some expressions are universal and automatic (Yang, T. et al., 2020). Based on the activation theory, emotions play several roles that are relevant to aspects of design (Fitch & Tarbutt, 2016). Some scholars denote the 'emotional experience' to explain that a design serves as a tool for designers to deliver ideas and emotions (Keen, 2015). Emotion is not only a mood descriptor but also a behavioural indicator (Desmet, 2015).

Emotion is a central quality of human existence, and most of our behaviours, motivations and thoughts are enriched with and influenced by emotions (Ekman, 2003). The relationship between the world and humans is essentially affective, which means that all our interactions imply and involve emotions, whether these interactions are with the social or the material world. Emotion can be classified as an affective state with moods, sentiments and personality traits. We can distinguish between traits and states, for example, dispositional attitudes towards something and short-term reactions. It has been suggested that the emotional response to design outcomes stems from the relation between how the design outcome is perceived and the user's concerns, which is true for emotions and attitudes.

Based on numerous design projects and studies, Norman (2004) argued that designers make decisions to solve problems when emotions work in tandem with cognition. Without emotions, making decisions is challenging. Emotions evaluate the environment and provide insight. Neurochemicals in the brain influence perception, decision-making and behaviour as a result of emotional states. There are three levels of design:

1. visceral (the automatic, pre-wired layer);
2. behavioural (the parts of the brain and its processes that control everyday behaviour);
3. reflective (the contemplative part of the brain).

Each of the three design levels above contributes to the formation of user experience. The visceral level makes quick decisions, sends the proper signals to the muscles (the motor system) and notifies other areas of the brain. This is the beginning of emotive processing, which is determined physiologically and may be blocked or augmented by control signals from the above three levels of design. The behavioural level includes the majority of human activity; actions at this level may be amplified or inhibited by the reflective layer, which can enhance or inhibit the visceral layer. The superior layer is reflective thinking. Notably, it does not have direct access to either sensory input or behavioural control. Instead, it monitors, reflects on and attempts to influence the behavioural level.

Benedict Singleton (2025) noted that design is not an experience that can be created. Emotions in relation to design outcomes may be comparatively weak because design outcomes do not always have a large impact on what we do. In many cases, it is not the design outcome per se that is important to users but whether it plays a central role in their activities. Emotions stem from an appraisal in which the meaning of a situation is compared to the needs and motives (concerns) of users (Salas et al., 2012; Rodrigues et al., 2021). To understand the contribution of a design outcome to emotions, we must understand what role the user thinks it plays in a situation. While this may to some extent relate to a design outcome's expression, it can in many cases relate to the design outcome's functionality and behaviour, that is, how well it mediates an activity.

Using theories of situated cognition, Rogoff and Lave (1984) investigated cognitive thinking processes in creative education and the way they impact learning. The researchers invited a group of students who were well-versed in these processes and served as a good sample. Their turn-taking order was established at the start of the task and maintained throughout the entire process. During each turn, one student was in charge of mixing chemicals and recording the outcomes. This does not mean that the students worked in isolation; several decisions about what to combine and how to define the results were reached after extensive discussion. At the start of each new turn, one student selected alternative chemicals, and the other children compared them to the record. There were six possible pairs to choose from but no apparent order in which they could be presented. Up to the last two pairs, the students had no trouble selecting a new pair that had not already been chosen. There was no apparent system in place to arrive at the final two pairs, but there was growing concern about the prospect of finding more work.

Design scholars generally believe that the act of designing involves cognitive processes. However, the close relationships between cognitive processes and emotions have seldom been recognised. Before exploring the nature of emotion and its role in the design process, it is essential to observe the relationships between emotion and cognitive processes, emotional reactions and responses. Using a psychological theoretical framework, emotion can be described as a complex combination of subjective and objective interactions governed by hormone systems in the brain. These hormone systems influence affective experiences, such as arousal, pleasure and displeasure, as well as cognitive processes, such as underlying impacts on personality and perception. This often leads to communicative, goal-directed social competence (Azadi et al., 2019). This concept emphasises the variety of factors involved in emotions and includes four distinct elements that may comprise an emotion: the individual experience, the associated cognitive processes, the physiological reactions and the behavioural response. It is crucial to note that any assessment of the experience and cognitive functions associated with emotion is primarily subjective; that is, only the individual experiencing the emotion can assess and rate it. However, physiological signals, such as those that can be recorded by electrodermal activity measuring tools (Egger et al., 2019) and brain wave measuring systems (Nunez & Srinivasan, 2006), which provide measurable markers for assessment, may be used to quantify the physiological changes that emotions produce. The expression of a resulting behaviour is likewise observable and quantifiable.

An emotional response has three different physiological aspects that relate to behaviour and autonomic and hormonal activities. Different brain systems regulate these components. It has been demonstrated that when emotionally charged inputs are presented, single neurons in several amygdala nuclei in the brain are activated (Pessoa, 2022). The lateral, basal and central nuclei are the elements of the amygdala that receive emotional inputs from all areas of the neocortex. The amygdala's primary nucleus is the brain's most important emotional reaction centre. Its activation happens in response to frightening stimuli, and the absence of an emotional response when this happens is due to the inactivation of these regions (Bucci et al., 2021). Emotional responses are generated by the central nucleus, which sends projections to other brain regions (the hypothalamus, midbrain, pons and medulla). The hypothalamus (Giotakos, 2022) controls autonomic circuits to mobilise the proper physiological responses for specific emotional states. Different activation patterns characterise various events or experiences and their corresponding emotional states. For instance, a stressful event or threatening circumstance can cause fear, rage or grief, which emerge autonomically. When individuals are happy, their physiology responds differently. Heart rate (Pilcher et al., 2022), blood flow (Smith & Fernhall, 2022) perspiration (Tashjian et al., 2022) and gastrointestinal motility may all be affected by variations in the response of the visceral autonomic motor systems. These are produced by the central autonomic network, which combines sympathetic and parasympathetic responses. Complex and distinctive forebrain-mediated stimuli, such as films, books and music, can also stimulate autonomic and sympathetic responses.

Emotions are linked to the amygdala; consequently, psychophysiological tests that rely on the autonomic nervous system can provide accurate readings of the physiological aspects of emotions (Vesisenaho et al., 2019). Several scholars noted that designers' understanding of the environment around them influences their emotional reactions. During the design process, designers focus primarily on the logical thinking process to provide solutions to design problems rather than on the emotional changes. Therefore, the influence of emotional responses may be undervalued. Few designers understand that emotional fluctuations can affect decision-making and, accordingly, influence design outcomes.

2.5 DESIGN AND EDUCATION

2.5.1 Design Process in Education

Design teaching and learning emphasise sustained investigations and creating alternative viewpoints from typical meanings by using the available resources. Design students are expected to equip themselves with the ability to collaborate and explore solutions to complex design problems. Design education not only aims to inspire students to create their design outcomes but also cultivates the cultural spirit in which the design students live, thereby facilitating the development of their thinking as designers. In addition to the general approaches to literacy education, there is a trend towards student-centred pedagogies that motivate students to actively participate in design communities.

Knowing how to develop knowledge can be learned using several methods, all of which are valuable. Several scholars emphasised the value of enquiry and the application of design outcomes as well as design planning and human-centred design (representing symbolic and real dimensions). It is crucial to have the capacity to utilise and produce knowledge. Lum (1997) explored the process through which junior design students use the findings of a study they previously conducted.

Schön (1987) pointed out that design studios provide a location for professional training in a broad variety of design disciplines. The research participants were instructed to use the concept of 'learning by doing' in the design studio. By working on projects that were comparable to professional activities, students were expected to study, think and behave like professionals. The approach emphasised learning by doing. The design studio was regarded as a workshop for professional practices. Schön recorded one of his students' learning-by-doing experiences and revealed that design educators have to create projects of appropriate size and arrangement to match the classroom unit, as these are essential considerations to reach a scale similar to real-world design projects.

Capps and Crawford (2013) suggested that enquiry-based learning may be a technique worth developing further. The participating students started by accumulating observations and then investigated why they had occurred. Next, they developed hypotheses to account for these findings and devised and carried out experiments to evaluate their ideas. The study was conducted in early August at a university and a natural history museum in the northeastern United States of America. It lasted approximately 60 hours. The main objective of the setting was to provide a realistic atmosphere for instructors to improve their comprehension of scientific knowledge, including evolution, geology and enquiry. Additionally, instructors were assisted in acquiring scientific methods and refining their pedagogical practices towards an enquiry-based approach. Teachers participated as learners in the Fossil Finders' curriculum and materials, attended many palaeontology/geology field excursions and gathered and analysed data as part of scientific research. Additionally, instructors were encouraged to think critically about their experiences and past teaching.

Rowland (2000) added that enquiry-based learning is a highly deliberate approach accessible to academics from all fields, regardless of their educational experience. While it is appropriate for both new and seasoned university teachers, it covers a vision of teaching and learning that questions several widely held assumptions about university teaching and its connection to disciplinary research. The *Enquiring University Teacher* is an engrossing portrayal of a professional development strategy in which one's own teaching serves as a stimulating topic of investigation. It emphasises the inherent value of learning about university teaching alongside colleagues who provide perspectives from a variety of fields, resulting in nuanced knowledge about teaching and learning processes. The book delves into the essence of the university teacher's research process as a collaborative and personally reflective mode of professional development. It entails challenging personal and intellectual ideals and integrating them into academic instruction.

Using the example of enquiry-based learning as a process of investigating information and as a way of teaching, Arias and Davis (2017) expressed their point of view on the subject. Their study presents three critical characteristics that influence the engagement of creative students. The research employed a mixed-methods design that included multiple methods of data collection and analysis. A study conducted by Sari et al. (2020) examined the impact of enquiry-based teaching on students' creative thinking skills, which was one of the variables that was categorised into higher and lower levels. Creative thinking was found to act as a moderator between high and low levels of creativity. A factorial design was used to investigate the main effects of the two experimental variables and the interactions between them. Sari's findings reflect the need to teach and guide creativity to enhance students' creative thinking skills.

2.5.1.1 Complex Cognitive Thinking in Design

The experienced designer's ability to navigate the design process in a seemingly effortless fashion is the result of human cognitive architecture. Theories that support this include Wegner's (2002) dual-process model of cognition, which posits the configuration of human cognition (Martinez, 2010). Scholars have investigated two types of memory that work together, namely working memory (Sweller et al., 1998) and long-term memory (Van Merriënboer & Sweller, 2005). Working memory processes input and connects new knowledge to long-term stored knowledge, creating connected webs of knowledge called *schema*. Experienced designers have large, highly connected schemata that enable them to work efficiently and automatically by incorporating large amounts of highly connected knowledge into their limited working memory. This is particularly true of highly complex domains such as designing for the elderly or physically handicapped persons (Van Dooren et al., 2014). The effort needed to process non-automatic schema in the working memory is called *cognitive thinking* and plays a critical role in determining to what extent novices studying complex domains can learn effectively (Feldon, 2007). Junior design students lack organised schemata and instead use trial-and-error methods that quickly fill the working memory, leaving them with little capacity to make new schema connections (Van Merriënboer & Sweller, 2005). Sweller (1998) proposed three categories of cognitive thinking: intrinsic, germane and extraneous thinking. *Intrinsic thinking* relates to the inherent complexity of a task; complex tasks such as designing for older adults require high intrinsic thinking. *Germane thinking* is needed for the brain to make new schema connections. *Extraneous thinking* constitutes distractions or thinking that takes away from the capacity to build schema or handle innate task complexity. Thus, in complex domains such as design, highly developed cognitive thinking can interfere with learning by filling up working memory and leaving no capacity to build new schema connections (Van Dooren et al., 2014; Van Merrienboer & Sweller, 2005).

2.5.1.2 Cognitive Thinking and Emotions

In the case of intrinsic thinking, the addition of affective skills adds to the complexity of a task. In design studies, junior students are expected to employ empathy to understand what it feels like to be a user of the space or product they are designing and use the gained perspective to select appropriate design choices (Altay et al., 2016; Walther et al., 2017). Empathy is a learnable skill, and junior design students have to recognise that the user's perspective is valuable and must choose to engage empathetically before they can design effectively (Zoltowski et al., 2012). Such engagement involves not only germane thinking but also highly complex learning and consideration skills during the design process. Thus, the development of instructional means to help junior design students build affective skills in design is important to not only maximise the potential for cognitive learning but also develop the skills needed to make appropriate design decisions.

Multiple researchers (Lazarus, 1984; Schachter & Singer, 1962; Zajonc, 1998) have posited a strong connection between emotional experiences and the cognitive appraisal of those experiences; however, the precise role of specific emotions during the learning process is still unclear (Park et al., 2015). For example, some researchers (Xue et al., 2020) indicated that negative and positive moods induced by emotional experiences can influence students' cognitive processing by affecting their ability to regulate abstract thinking and problem-solving strategy selection. Desmet (2018) proposed that an individual experiences a variety of emotions, including fear, amusement, rage, pleasure, regret and desire. Compared to a single feeling, a blend of different emotions can give rise to an enjoyable experience. The same likely holds true for other instances of enjoyment, such as sharing a joke, using products or communicating with a machine. Other researchers identified the counterintuitive and positive effects of negative emotions such as confusion (D'Mello et al., 2014). These effects are usually regarded as elements of psychological disciplines rather than design disciplines. Few design researchers engage in developing these areas and applying them as cross-disciplinary studies involving both design and emotion. Fried and Chapman (2012) regarded emotional reaction and inclusive human concerns as paramount to design. As individuals become increasingly sensitive to the dimensions of things that transcend the conventional characteristics of 'usefulness', the need to understand and establish emotional and aesthetic connections between people and products continues to grow. Similarly, the idea of experiencing where the subject and object meet and integrate is a crucial factor in the development of emotionally significant products. This is because experience is a space in which all faculties, particularly emotions, are aroused. This corresponds to Freud's (2012) theory of attachment between people and objects, in which, at the height of admiration, the border between the personality and the object threatens to dissolve. More work is needed to explore how emotions affect the way junior design students learn to productively respond to emotional experiences and which aspects of learning interventions induce effective emotional experiences.

Some academic research on design studies relied even more strongly on psychological concepts (Demir et al., 2009; Markussen, 2017). Enhanced concepts concerning how designers might implement the dominant assessment theory related to the psychology of emotion have been investigated. By analysing the distinctive evaluation components in people's accounts of emotional experiences with items, scholars have attempted to develop design-specific assessment structures (i.e., appraisal theory) for emotions such as pleasure, happiness and disappointment. Markussen (2017) utilised the appraisal theory and 'blended' it with embodied cognition, a major theory in cognitive science, to explain why patients have such a broad variety of views on using interactive healthcare systems. Markussen's theoretically informed research of interaction with a blood-drawing robot provides insight into how consumers produce new evaluations of objects (via the blending of conceptual structures) that evoke new or different emotional responses than those elicited by each individual, unblended structure.

In addition, design scholars have examined the cognitive and social dimensions of writing and other literate activities. These studies present guided intervention as a method of assisting cognitive development in junior students, which is a collaborative process based on shared experiences and knowledge. The goal of this strategy is for junior students to acquire critical thinking skills and subject-related abilities via directed social engagement with adults. The senior students model a set of behaviours, coach the junior students in the activity and help them through the reflective process involving the new experience to master the new behaviours. Thus, the students not only learn a new skill but also gain knowledge of the skill that the senior individuals had acquired through experience. Awang and Ramly (2008) studied how problem-based learning pedagogy and practice can be used in a design classroom and found that the feedback supplied by community members as part of the social cooperation process, as well as the perceived consequences of this input on learners, are both significant factors that can help junior design students broaden their approach to their design learning.

Rodrigues et al. (2019) explored the relationships between emotional intelligence (EI) and creativity. The findings of their study show that personal attitudes towards entrepreneurship and personal branding have a positive impact on entrepreneurial intention and that personal branding has a mediating effect on the impact of EI and its role in entrepreneurial intention. Accordingly, EI has a direct and beneficial impact on one's ability to be creative.

Bennett (2014) advocated for advanced abilities in problem-solving, critical thinking and analysing to be taught in formal settings. Those who participated in the research received instruction in advanced problem-solving, critical thinking and analytical abilities. Bennett's proposal comprised a condensed version of the work of earlier educators. The results show that increasing students' skills in planning and computer use will enhance their creative learning outcomes. Although problem-based learning has long been an established approach, its use in a rich media learning environment is still in its early stages. Problem-based learning is a teaching strategy used to provide recommendations to students to enable them to build up their learning abilities from their daily experiences. Using a similar approach, Kozbelt (2019) further developed evolutionary approaches to creativity. They argued that the capacity for human creativity can be seen as a design outcome of evolution.

The long-term ecological sustainability of human creativity is one of the intriguing elements of the criterion for measuring the worth of creativity. The evolutionary process itself improves over time, which led Kozbelt (2019) to research whether the creative process had historically been invariant (Dawkins & Furnham, 1989). Biological evolution, creativity, the mediating role of culture and the links between them and co-evolution have begun to provide significant answers to basic human concerns. Moreover, they have provided a level of control over our own development that is unmatched in the industry. The evolutionary process has shaped human creative abilities from daily experiences. Based on adaptive behaviours, humans have developed the most remarkable set of creative abilities (Ciompi & Panksepp, 2005).

Clinton and Reiber (2010) explored how instructional design and development are used in studio-based learning. As part of their investigation, they examined the studio curriculum. The research was inspired by a design and technology programme at a well-known research-driven institution that exemplified the careful application of ideas about how today's students learn with complicated information. In 2009, Brocato proposed a theoretical framework to govern studio-based learning, an educational approach that focuses on hands-on application, thorough exploration, and innovative thinking inside a regulated studio environment. This practice is particularly widespread in artistic, architectural, design and other creative domains. Brocato recognised that the academic research illuminates the fundamental aspects of studio-based learning, offering useful insights for effectively managing classroom settings that promote student-centred learning. By understanding and applying these essential elements, teachers may effectively create learning environments characterised by student-centred approaches, as noted by Brocato. The studio-based approach offers useful insights into addressing challenges in classroom management by prioritising the creation of conducive conditions that promote active engagement, creative problem-solving and self-directed learning among students. This instructional paradigm creates a framework for fostering an engaging and supportive learning environment, characterised by individualised supervision, collaborative work and active participation. These are essential elements for effectively managing studio-based learning environments. As research continues to improve our understanding of studio-based learning, there is a need to explore how pedagogy may be utilised in management courses in creative and design studies.

Greenberg's (2009) research explicitly concerned the innovation and critique ability of junior design students. In practice, students typically reflect enjoyment in completing their projects and professional portfolios. Many are highly imaginative in terms of developing their portfolios, to the point where the portfolio itself exhibits the student's talents. Many students put substantial care into creating their professional portfolios, using high-quality materials and expanding them to incorporate initiatives beyond their courses. Many of the study's participants mentioned that they used the portfolios during real job interviews and thought it made a difference in setting them apart from other candidates. The study's conclusion confirmed this.

Heresco-Levy et al. (1990) investigated the means of teaching design using Bauhaus principles as a baseline. They examined what it means to teach design and battle with concerns of what designers need to know. At the time, design fields began to re-articulate the role of design in terms of the constructive, non-neutral and recursive axioms of the emerging systemic paradigm. For example, they developed instrumental and existential methodological coherence in terms of the cognition, creation and design outcome of artefacts. In addition, a desire to understand the existential characteristics of the *being* of designed artefacts arose. In this knowledge-based context, the designers' learning process underwent fundamental changes that had far-reaching effects on design practice, design responsibility and design credibility.

Brew (2012) examined the way communities involve teachers and staff as co-learners. There is a need to rethink the function of higher education and to renegotiate teacher–student interaction. Based on changes in perspectives, a new paradigm is required that places a premium on student-centred education, conceptual transformation and a view of knowledge as being formed within a socio-political context. The new model incorporates Lave and Wenger's (1999) ideas about situated learning, Wood's concept in Boyer Commission on Educating Undergraduates in the Research University (2003) recommendations for redefining the nature of academic work and undergraduate education, Prosser and Trigwell's (1999) work on teaching conceptions and my aforementioned work on knowledge, research and scholarship conceptions. Brew (2012) offered a novel, although rather romantic, picture of higher education that compels us to reconsider its essence.

Several scholars emphasised the role of emotion and inclusive concepts. Le Heron (2013) examined the extent to which communities engage teachers and 'employees in the design industry as co-learners. Students, professors, community members and external contributors/scene setters may all benefit significantly from working together, particularly if fieldwork is designed and organised in a manner that maximises co-learning opportunities. Everyone engaged in fieldwork is, in some way, a co-learner. Students and faculty members may all benefit from listening to or talking with local experts who describe or demonstrate how knowledge about a location, interaction, phenomenon or consequence is generated. Students may benefit from one another and co-learners if opportunities are available for them to share insights into their diverse worldviews, sociocultural perspectives and past learning experiences.

Design teaching and learning emphasise sustained investigating and creating alternative viewpoints from typical understandings through resource organisation. Design students are expected to equip themselves with the ability to collaborate and explore solutions to complex design problems (Liem & Sigurjonsson, 2011). Design education not only inspires students to generate design outcomes but also cultivates the cultural spirit in which they work, thereby facilitating the development of their thinking as designers (Squire, 2008). In addition to the general approaches to literacy education, the recognised trend towards student-centred pedagogies motivates students to actively participate in design communities.

Design educators such as Sahinkaya and Burrows (1985) emphasised knowledge construction in design. They claimed that design knowledge and skills should be acquired systematically. Design students learn from contextual problems arising from learning materials and through the support of their instructors. Studies on design education have suggested studio-based pedagogy as a method for cultivating the abilities of design students. This method prompts students to develop their conceptual understanding of design and the design process and supports their thinking as designers (Kuhn, 1998, pp. 65–71; Schön, 1987). Mathews (2010, pp. 87–102) agreed with this approach and investigated how educators can prompt design students to think like designers. Their research emphasised the importance of the studio setting that educators prepare and encouraged educators to combine project-based work with open-ended problems. The core of problem-based learning is collaboration and personal reflection (Lebbon et al., 2009, pp. 159–167; Restrepo & Christiaans, 2004, pp. 123–138). In addition to these design teaching and learning methods, educators have investigated how new trends in design education should be developed (Wormald & Rodber, 2008, pp. 294–303). James (2017) suggested that before designers begin problem-solving, they must understand the nature of the problem. This understanding is regarded as self-awareness and is an essential ability for designers, as it enables them to make a positive start to their design projects. Berman and his colleagues (2020, pp. 127–140) used a similar concept regarding the importance of designers' understanding of the nature of a problem. They expanded the scope of this consideration by stating that social aspects no longer provide a solution to a design problem but constitute strategic communication tools. Bixler (2016, pp. 273–285) supported the shift of the leading role in teaching and learning from the teacher to the students. As a result of this shift, learner motivation has become a major concern in the development, monitoring and assessment of instructional effectiveness in the teaching and learning process. Cooper et al. (2013) indicated that designers' self-awareness and self-management are abilities used to manage conflict. However, designers tend not to manage design effectively. Salama (2009, pp. 8–27) proposed that contemporary designers must confront and respond to the challenge of creating an environment that supports, enriches and celebrates human activities. The ability to do so results from a combination of social, cultural and economic needs. These demands require designers to assume responsibilities and subsequently make decisions regarding these responsibilities. Owen (1998) described the nature of design learning as a cross-disciplinary subject. Design was regarded as a special subject, and its nature was found to be different from that of art and science. Design learning developed its own particular purposes, significance, measurements and learning processes, which supported the development of design learning. However, limited investigation, discussion, development and theoretical research followed, resulting in a limited understanding of developing theoretical design learning with a knowledgeable foundation.

2.5.2 Five Main Attitudes in the Design Learning Process

Miles et al. (2014) investigated research methods to obtain quantitative and qualitative data and proposed that a complete paradigm for the management, analysis and interpretation of qualitative empirical data be developed. While emphasising the relationship between a theory, its ideas and its empirical indicators, their approach differentiated between within- and between-case analyses. The theory's ideas were mirrored in the actual data. The results demonstrated that researchers may proceed from field notes to create conceptual knowledge of the studied processes, using codes, memoranda and diagrams. This iterative technique is completely open to discovery, and the comparative model emphasises the handling of negative instances.

The concept of saturation provides guidelines for sample size in qualitative research (Mason, 2010). In this study, a categorisation of 26 distinct qualitative research techniques from different fields was used to examine any discrepancies between various research methodologies (Tesch, 1990). Twenty-six distinct forms of qualitative methodological tradition were discovered while researching qualitative research software. They were divided into four groups: linguistic features, regularity findings, meaning understanding and reflection.

In the initial investigation of the trends in design education, five main attitudes in the design learning process were identified. A review of studies conducted by psychologists revealed that emotion is linked to the five attitudes. Specifically, emotion influences appraisal and judgement in decision-making. Stimuli from the external environment create emotional responses and guide the thinking process over time (Atkinson & Shiffrin, 1968). This thinking process involves motivation and appraisal to support judgement and responses (Craik & Lockhart, 1972, pp. 671–684; Longueville & Gardoni, 2003). Scherer (2005) described emotion as a sequence of reactions towards external stimuli that shape a person's perspectives. Following an event, a person's intrinsic pleasantness and goal relevance emanate appraisal. This subsequently affects the person's ability to control their responses. Ho and Chau (2016, pp. 13–24) followed Scherer's study and observed that appraisal generates a result as an attribution of causation, which directs a person's responses, including decision-making and actions. Furthermore, studies on EI identified self-awareness, social awareness, motivation, self-management and responsible decision-making as skills that enhance emotional management. To explain the relationships between emotion and the five main attitudes in design education, the model presented in Figure 2.1 was proposed.

2.6 EMOTIONAL INTELLIGENCE

2.6.1 Emotional Intelligence Training

Mattingly and Kraiger (2019) investigated the probable route to enhanced EI training. The findings of their meta-analysis add both theoretical and practical advances to the EI literature by determining the degree to which EI can be taught. The modest and favourable impact of training on EI supports the flexibility of this construct, allowing us to conclude that EI is trainable. Studies were collected by conducting a computerised search for the words and phrases 'emotional intelligence', 'training' and 'intervention' using databases such as Business Source Premier, Dissertation & Theses Global and Google Scholar. Mattingly and Kraiger also identified relevant research in articles that were discovered during this search.

Some user experience research reported that user feedback on design outcomes and design conception indicates that users interpret their expectations. The scholar noted that this point of view is not only supported by the qualitative feedback from

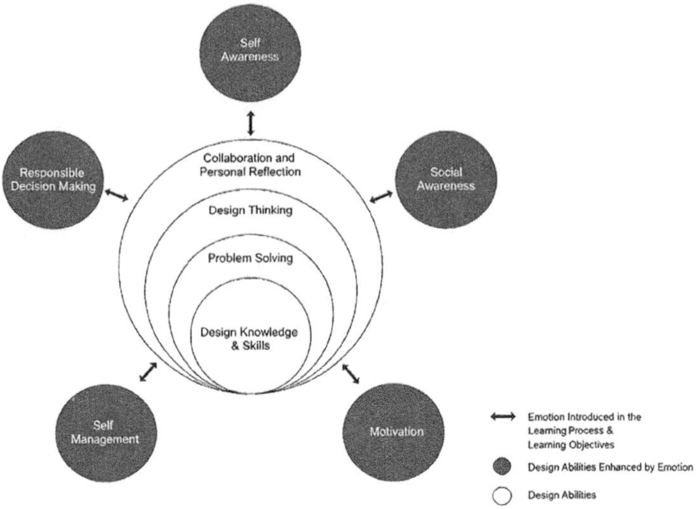

FIGURE 2.1 A model illustrating five main attitudes in design education

design experience questionnaires but can also be explained by the physical mesolimbic dopamine system of the human body (Tzvetanova et al., 2007; Norman, 2017). The 'mesolimbic dopamine system', theorised by the somatic-marker hypothesis, is an important aspect of unconscious action bias. The mesolimbic dopamine system has increasingly become a topic of discussion in terms of playing a role in reward processing and decision-making. Participants were invited to present their responses to stimuli (Longueville & Gardoni, 2003). Those responses were recorded and compared with the predicted primary responses. The participants' feedback showed that their responses shifted over time from occurring while receiving primary reinforcement to occurring and being reinforced at the onset of predictive cues. By employing sophisticated computational techniques to shape the model, activity within the mesolimbic dopamine system signals was found to cause biased behaviour between actual and predicted rewards.

2.6.2 Emotional Literacy

The formation of EI should not be overlooked. Teachers should pay careful attention to the emotions of learners. Instructions are vital to aid and direct learners towards developing EI and attainments (Huanhuan, 2016).

Emotions affect students' advances from various angles, and emotional education is inseparable from education as a whole. This study focuses on teachers' function as emotional guides in 'flipped classroom' teaching situations. Grounded in a comprehensive assessment of emotional education and an analysis of the teacher emotion guides, the research team provided solutions for teachers to enhance students' academic achievements and EI. It is proposed that teachers' role as emotional guides in flipped classroom teaching environments may be carried out on two levels, namely guidance to prevent negative emotions in learners and guidance to generate positive emotions in learners. It is projected that teachers' position as emotional guides should establish long-term learning results and promote emotional development in a flipped classroom teaching context.

Emotional skills development can be included in a naturalistic and implicit teaching environment. It is possible to enhance students' learning outcomes, foster prosocial conduct and support healthy emotional development by fostering the development of emotional skills in elementary (primary) classrooms. The emotional climate of the classroom refers to the naturalistic and implicit teaching environment that incorporates emotional skills development. This research focused on the emotional content of spontaneous teacher–student interactions. Researchers examined how instructors manage emotional situations and the positive tactics they use to do so. During 60 hours of classroom observation, six instructors who had been nominated for their remarkably positive classroom settings were observed. As a result of the observations, the following themes emerged: (1) developing classroom connections, (2) creating and regulating emotional guidelines, (3) being emotionally aware and (4) managing emotional circumstances. Harvey et al.'s (2012) model of the emotional atmosphere of the classroom was supported by this research. The study results laid the groundwork for the future development of emotional norms.

The importance of cultivating EI should not be underestimated, and instructors must pay special attention to the various emotions reflected by learners (Huanhuan & Chi, 2016). Instructions are essential to guide learners towards developing EI, and the relevant emotional guidelines should be used as a tool to enhance junior design students' emotional literacy.

Fostering emotional skills in the elementary (primary) school classroom can lead to improved learning outcomes, prosocial behaviour and positive emotional development. Incorporating emotional skills development into a naturalistic and implicit teaching environment is a key feature of what is meant by 'the emotional climate of the classroom'. The emotional content of spontaneous teacher–student interactions was of interest in this study. The investigation focused on how teachers manage emotional events and, in particular, what positive strategies they use to do so.

2.7 DESIGN AND EMOTION CONCEPT IMPLEMENTATION IN THE DESIGN PROCESS

2.7.1 Roles and Functions of Emotion in the Design Process

Norman (2017) concluded that people are often motivated to develop emotional recollections and connect their feelings to their sensory experiences via different forms of stimulation, as well as other types of stimulation, including the visual and tactile properties of design output. Due to the wide range of potential ways to stimulate and identify emotional responses through design experience, it can be challenging to define and implement consistent approaches to design-related stimulation and to accurately recognise emotional responses to design-based stimulation. Therefore, it is important for design researchers to continue exploring and developing methods to better understand emotional responses through interdisciplined research. There are examples of specific stimulation approaches that may facilitate the recognition of emotional responses through design experience, such as eye movement tracking and clickstream data that are utilised in bio-science research studies. Therefore, researchers studying the effects of stimulation on emotional responses through design experience should explore empirical methods to quantify and study the effects of design-related stimulation on emotional responses, which will ultimately aid in interdisciplinary research on emotional responses to design experience. Tzvetanova et al. (2007) studied the influence of user/designer emotions on design outcomes and provided a method to analyse the emotional responses of end-users/audiences to design. Data from their study provide insight into how the industry measures the emotional connections between designers/users/customers and design outcomes. Although their study provides valuable insight into the emotional impact of design on users/customers, it does not illustrate the emotional impact of the design process on the designers/end-users/audience. As stated in activism theories, emotions play a role in design, and according to Badran & Al-Haddad (2018) many companies are attempting to systematically study design processes to identify problems early on in the design process. Design Process Model seeks to address the problem of the effects of emotions on various stages of the design process, including idea generation, design definition and evaluation. Five major categories of user experience were identified in Badran and Al-Haddad's study, including utility, usability, aesthetics, identification and value.

Wang and Zhou (2020) conducted research on emotional product design concepts. Consumers have experienced increased individualisation and diversification in their personal tastes, consumption levels and aesthetic standards. The advancements in the industrial and technological sectors have enabled consumers to realise functional results while meeting their emotional needs. Wang and Zhou proposed a method to establish design decisions based on the user's Kansei descriptions. EGM expert interviews were used to create a database of user images and product form attributes. The perceptual image components that yielded the highest user satisfaction levels were determined and used in the programme. The program then calculated the adaptive assessment of user preferences by progressively evaluating individuals based on user evaluations. This method not only facilitated user decision-making during product development but also enhanced the accuracy of user evaluations. Additionally, the product development process incorporated consumer preferences. The fuzzy Kano model (Wang & Zhou, 2020) was applied to increase the scope of product modelling evaluation. Violante et al. (2019) studied the application of emotional design and virtual reality to develop research results. A future consumer may interact with such a product and comment on its appeal, texture, functionality, and their emotional response to the product (Violante et al., 2019). Consequently, new approaches to designing user-centred experience design are being developed and creating an 'emotional design' process. Further examination of the user-centred experience appears to be best achieved through the study of facial expressions, which appear to be the most reliable and attractive method of doing so. Innovations in information technologies are now widespread across many industries and areas of application. Virtual reality and related products are increasingly popular and represent one area of information technology innovation. When products exist in a virtual format, product creation and development require more efficient management of resources than when a product exists in a non-virtual format. Regarding resource utilisation and sustainable development, the primary advantage of virtual products is the ability to produce products early in the development process, even during the conceptual phase. This represents an opportunity for designers to explore the boundaries of user-centred experience design approaches.

Friedman (2010) emphasised the emotional characteristics of product design. He explored a brief history of emotion, emotional design principles and design thinking. Friedman's findings indicated that customer emotional responses to products significantly influence how they perceive, approach and utilise products. Designers are increasingly paying attention to the emotional elements in product design. Emotional elements and the emotional elements of product design are essential for designers to understand (Friedman, 2010). Through the study of pioneers and their successes in design, designers can learn to categorise emotions, isolate emotional elements and apply emotional theory to their own work. Systematic emotional design approaches and procedures should be detailed and discussed among design experts. Many products and services are designed without regard for individualised needs of consumers; therefore, there is a greater likelihood of error and miscommunication (Marin-Alvarez et al., 2020).

As discovered by design professionals, instructions for designs that are not easily understood can lead to frustration, confusion and excessive fatigue for users who are obligated to follow them. To comprehend the type of instruction needed to provide adequate guidance for users, along with the specifics of the design features and specifications, it is essential for designers to understand the individual who will be utilising the design outcome. To classify users appropriately, designers must take into consideration their specific preferences, limitations and needs. Unlike the literature, relatively few products and services are based on the needs of a single user, thus providing a high likelihood of occurrence of errors and misunderstandings. Formosa et al. (2025) focused on enhancing user experience with tangible outcomes and measurement tools. Clinical products are referenced as most un-certified designs contain some instructional content that confuses many consumers. The designers did not take into consideration the other users, such as patients and/or family members, which resulted in the consumers being unable to understand the information that was provided to them (Formosa, 2019). A classification system for the various users that interact with clinical products must be developed to provide solutions that satisfy their cognitive requirements. Formosa's study illustrated the importance of designers' objectives, their ability to achieve those objectives, and the amount of effort that they expend in pursuit of those objectives, as all three are essential components of designers' motivation in a specific circumstance (i.e., the concept is defined as self-motivation). As a result of Formosa's findings, Boess et al. (2009) investigated designers' knowledge of consumers' needs and how it might help to clarify the problem and optimise the design process. Designers' methods to design and evaluate their designs, using product usage data to create products that meet consumers' needs, were also examined in Boess et al.'s (2009) study. Their study illustrated that consumers expect products and services to perform based on their unique background factors, such as geography, buyer type, and gender, both statistically and monitored. Therefore, Boess et al. (2009) study provided the impetus for examining the concepts of empathy to understand consumers and designers as needing to develop empathic understanding in addition to their EI abilities.

2.7.2 Emotional Typology

There is a need for improved systematic training that provides a general understanding of emotion, particularly in the design process. Some designers need to gain experience in managing their emotions. They may be unable to identify their emotional changes accurately (Heskett, 2017), describing them as positive or negative or with generic terms such as 'happiness', 'joy', and 'anger'. This terminology is described as layman's language and is insufficient for designers to identify users' emotional changes. Hence, it is essential to have a general understanding of the categorisation of emotions to develop a solid understanding and description of design and emotion studies (Mastria et al., 2019).

The two main approaches to grouping emotions involve discrete and dimensional models. These two models suggest different methods of identifying emotions. The discrete model proposes that some emotions are relatively simple to identify (Harley et al., 2019). Each emotion comprises a distinct collection of experiences, physiology and related behaviours (Ekman, 1992; Le Roux et al., 2022). However, the classification of these fundamental emotions has generated some debate. Ekman (1992) postulated six primary emotions, namely happiness, contempt, rage, fear, surprise and sadness, that are reflected primarily via facial expressions. Plutchik (1994) suggested a model that includes trust and expectations. Other scholars asserted that simple emotions, such as shame, humiliation, remorse and pride, may be combined to create complex feelings. These feelings are known as self-conscious emotions, since they are informed by societal norms and acceptable behaviour (Lewis, 1995). The dimensional model analyses emotions in terms of psychological components (Russell, 1983), namely valence or polarity (Lang, 1988), as well as physical activation or arousal (Mauss & Robinson, 2009). Emotional experiences known as valence or polarity may be positive (pleasant feelings) or negative (unpleasant emotions). The physiological response that accompanies an emotional experience is known as physical activation or arousal and may range from low to intense activation. For example, the model suggests that fear is evaluated as negative and has a high level of physical activity, whereas sorrow is defined as negative (in terms of intensity) and involves a modest level of physical activity (Casals-Gutierrez & Abbey, 2020). The dimensional model enables the portrayal of several nuanced aspects of a specific discrete emotion. For instance, although the term 'happy' has positive connotations, it can also express a range of degrees of activity via words used to describe the concept of happiness, such as enthusiasm and joy.

Other scholars postulated new ideas, such as the counselling approach (Lang, 1988), emotional avoidance (Weiss et al., 2018) and emotion management (Mauss

& Robinson 2009). In contrast with positive emotions, negative emotions may be overpowering and evoke a sense of being controlled. Individuals have control over negative emotions when such feelings do not significantly alter their physical or mental condition. This is a feature that consumers can easily evaluate, and it is included in arbitrary emotional measurements (Desmet, 2018). The impulse to pursue (approach–avoidance behaviour) or avoid (withdrawal behaviour) the source of the feeling can be affirmed. There is no superior or inferior approach. One can simply employ the approach that provides access to a similar phenomenon from several perspectives. Regarding valence, the approach–avoidance method may be positive (elation) or negative (fear), while withdrawal behaviour can be viewed as positive (relief) or negative (anxiety/depression; Carver, 2006). It is challenging to include this component in objective measurements due to the subjectivity of emotion interpretation.

Based on these interpretations, scholars have summarised approach–avoidance and withdrawal behaviours in the dimensional emotion theory. Feelings such as sorrow, happiness and anger can be defined using dimensions of fear. Different methods can be used to measure emotions based on their combinations (usually negative and positive) and the body's response to them. In addition, scholars have classified emotions into three main dimensions, namely internal, external and natural, according to personal experiences recalled by the individual. Compared with normal emotions, preferential target emotions are generated by both internal and external processes (Harmon-Jones et al., 2017). Emotions under the external category are mostly elicited by stimuli obtained from the external environment (e.g., watching film clips). Emotions under the internal category are primarily elicited by initial personal events (e.g., recalling a personal experience). While scholars have combined the external and internal emotions (Harmon-Jones et al., 2017) as distinct primary emotions without contradicting the dimensional and discrete theories of emotions, the main categories of emotions are:

- natural;
- external negative;
- internal negative;
- internal positive; and
- external positive.

This categorisation of emotions is based on psychological theories and the emotions' influence on the decision-making abilities of designers, specifically those applied in the design process.

2.7.3 Emotional Granularity

Emotional granularity is a concept that was introduced by Hoemann et al. (2020), who define it as the 'specificity and complexity' with which we construct an instance of emotion based on the context and the relevant emotional characteristics.

Previous studies have been limited in capturing the intra-subjective variability of emotional granularity or the multiple dimensions of emotional granularity that may exist; for example, the relationship between various emotion terms and one another, because they utilised traditional within-subject correlation techniques that provided a single estimate per participant.

Network analysis was used to assess granularity of emotions on a situation-specific basis, by providing estimates of how well connected emotion terms are at any particular point in time. The method created network graphs with each node in the graph being an emotion term and each link between two nodes in the graph showing the nature of the relationship between them, enabling the researchers to understand the ways in which emotional experiences can change dynamically across time. The high-dimensional networks enable visualisation of the patterns of emotions, through grouping of highly interconnected nodes, thereby providing greater insight into group-level relationships of emotion. Person-specific networks were constructed using experience sampling methodology and the resultant networks were used to provide estimates of distinct emotion categories, complex interrelationships among these emotion categories, and changes in emotion categories, therefore allowing the researchers to obtain detailed assessments of the participants' momentary emotional experiences.

2.7.4 Emotional Literacy in the Design Process

A number of studies demonstrate that a variety of emotions are involved in the cognitive processes of young designers. In terms of design practice, negative emotions (e.g., sadness, anger) are likely to occur when a student's work is rejected, and positive emotions (e.g., joy, pride) are likely to occur when the student receives praise for his/her work. Students will also experience emotions outside of the design context (e.g., excitement from earning good grades, fear/stress from studying) but they are generally unaware of these emotional changes occurring during the design process and have little or no experience with managing their own emotions during the design process; thus, potentially inhibiting their ability to make progress. For example, a student who is anxious may not be able to draw upon his/her inner strength, and a student who suppresses anger may make decisions based on wounded pride rather than an objective assessment of what would be best.

Fear, while a very important emotion in human cognition (LeDoux, 2003; Li et al., 2012), has been shown to frequently go unnoticed by designers (Guerra & Tripp, 2018). Many designers believe that designers should not think too much about fear and its associated 'fear-based' thoughts and actions; instead, they should utilise knowledge and theory to minimise failure and manage potential threats more effectively. When designers recognise potential abstract threats (e.g., radiation hazards), they are able to maintain focus on decision-making. While fear can cause a designer to become stressed and can even lead to post-traumatic stress disorder (Dikmen-Yildiz et al., 2018); self-reflection enables designers to recognise their physical responses to fear (e.g., increase in heart rate) and recover their composure.

Although often overlooked as it is subtle, calmness represents the baseline emotional state for designers who are working in their comfort zone. The research by Horvath et al. (2018) indicates that when calm, the brain releases optimal amounts of dopamine and serotonin and activates the prefrontal cortex, which facilitates logical thought and sound judgement. When calm, designers can set aside their subjective feelings and make objective assessments of problems (Liang & Jin, 2011). Through the use of calm observation of the larger context (including long-term relationships with clients), designers develop the ability to manage their emotional fluctuations without losing their logic.

2.7.4.1 Typical Emotions in the Design Process

Research has found that many junior design students have difficulty understanding that there are eight primary emotions that are grouped together based on what users tell researchers about how they experience using a product or service (Herrmann et al., 2018). The eight primary emotions are elicited from outside of the user (i.e., the design itself); however, the designer's own emotions are generally the result of outside influences as well and therefore may not be an active component in the designer's thinking process, which shows a knowledge gap in previous design and emotion literature. A designer's emotions will arise as a response to the external stimuli that affect them and will influence how they make decisions. Users' emotions also create their reactions to the world around them in ways that differ from the way designers react to it. Studies utilising qualitative methods and appraisal theory (Jokinen & Silvennoinen, 2020; Laurans et al., 2009; Mahan, 2021) have indicated that users' emotions can affect the design process.

To identify usable emotions in design, the study group applied appraisal theory to understand basic emotions and compared those theoretical descriptions of basic emotions to both design empirical studies and practice. The review resulted in the identification of various types of emotions typically felt by a designer during the design process (see Table 2.1). Those emotions were then re-examined and incorporated into a new model representing relationships between emotions and common emotional experiences throughout the design process.

2.8 EMOTIONAL CONFIGURATION PRINCIPLES IN THE DESIGN PROCESS

Students need to be taught how their emotions affect their decisions and the outcome of their designs by developing an understanding of how emotions contribute to the design process (Ho & Siu, 2012). A conceptual model of emotions associated with design has been developed to demonstrate typical emotions associated with the design process (Oliver, 1993; Figure 2.2). The five categories of emotions are: external positive, internal positive, neutral, external negative and internal negative (Oliver, 1993).

Typical emotions are active emotions which drive motivation, judgement and action related to the design process, distinguishing them from moods. External

TABLE 2.1
Forty proposed typical emotions identified in the design process

No.	Emotion category	Emotion
1.	External negative	Nervousness
2.		Hatred
3.		Regret
4.		Discouragement
5.		Anger
6.		Jealousy
7.		Dissatisfaction
8.		Fear
9.	Internal negative	Sadness
10.		Frustration
11.		Loneliness
12.		Disgust
13.		Inferiority
14.		Anxiety
15.		Stress
16.		Longing
17.	Natural	Calmness
18.		Sympathy
19.		Affection
20.		Touched
21.		Satisfied
22.		Peacefulness
23.		Commitment
24.		Interest
25.	Internal positive	Pleasure
26.		Affinity
27.		Curiosity
28.		Pride
29.		Anticipation
30.		Respect
31.		Healing
32.		Gratefulness
33.	External positive	Joy
34.		Happiness
35.		Excitement
36.		Superiority
37.		Courage
38.		Relief
39.		Energetic
40.		Enjoyment

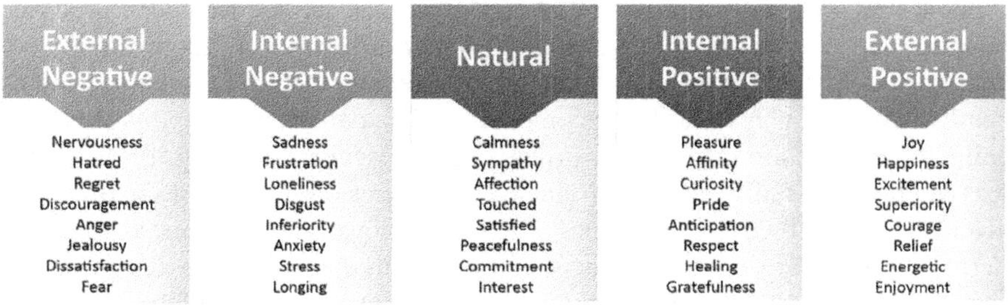

FIGURE 2.2 Forty emotion categories

control factors such as material allocation stimulate design based on external influences, whereas internal control factors are based upon the designer's own preferences and considerations. Neutral emotions result from a balance between the designer's perception of external influences and the designer's own reflections. Positive emotions (e.g., curiosity, satisfaction, love and pleasure) represent pleasant situational reactions. However, they are distinct from physical sensation.

Negative emotions (e.g., sadness, anger) can negatively impact the designer's confidence, self-esteem and well-being. Designers can manage and develop awareness of their emotions as they observe both themselves and their users, and use this knowledge to alter the way their emotions respond to each other. For example, the emotion of excitement (an external positive emotion) can become curiosity (an internal positive emotion), thereby encouraging the designer to investigate their subject in greater detail. Anxiety (an internal negative emotion) can be transformed into a peaceful (neutral) emotion, which is conducive to mental calmness and can either support or impede a design depending on how the designer manages their emotions.

Both emotion and cognition have a significant impact on the design process, and using balanced emotional responses will produce well-rounded design decisions. The proposed model of 40 typical emotions involved in the design process illustrates the relationship between emotion, thought and the emotional regulation of the designer, and provides a basis for future research into how designers experience, manage and utilise emotions throughout their professional activities.

2.8.1 Emotional Configuration Guidelines in the Design Process

2.8.1.1 Introducing Emotional Concerns into the Junior Designers' Design Process

Research regarding junior designers' views of theories of emotion indicates that emotion has a substantial effect on the design process (Ho & Siu, 2012). Researchers have also attempted to investigate if emotional knowledge can be learned (Mattingly & Kraiger, 2019), in order to determine the best method for increasing EI training through a meta-analysis. The findings of the meta-analysis add to the EI body of knowledge by determining the degree to which EI can be learned. The moderate and positive correlation between training and EI supports the malleability of this concept, so that design scholars can infer that EI can be learned. Junior designers must learn how emotion can affect their judgements, as well as the results of their designs. The majority of common emotions have strong involvement in decision-making and the cognitive thinking phases of the design process. The two primary factors that affect common emotions are information processing and decision-making. Both of these areas can motivate individuals and influence their decisions and actions in the future. Oliver (1993) identified five categories of basic emotions: neutral, externally positive, internally positive, externally negative and internally negative. The design process typically includes some external influences (such as material allocation) that emphasise external stimulation rather than junior designers' deliberation. Conversely, internal factors (such as the creator's personal preferences) prioritise human consideration over external stimuli. Neutral emotions represent relatively balanced relationships between internal and external stimuli. Positive emotions are different from pleasurable sensations and general positive effects; they include pleasant or desirable situational reactions, including interest and satisfaction to love and pleasure. Junior designers will be uncomfortable or depressed if they experience negative emotions. Negative emotions may lead to loss of confidence and self-esteem, a dislike for oneself and others, and feelings of overall negativity. The typical emotions experienced during the design process may vary or fluctuate based on the designer's experience, their prior research, and what they observe about themselves and their intended audience. For example, after being processed, curiosity, a positive internal emotion, could replace excitement, a pleasing external sensation. Therefore, junior designers can cultivate a robust curiosity about the unknown and the nature of things. Anxiety, a negative emotion that falls into the neutral category, can be broken down and changed into tranquillity, a state of being mentally calm. If junior designers find their minds burdened by worry and to-do lists, it would be useful for them to concentrate on calmer ideas. Changes can produce either better or worse design outcomes. Reasoning and emotion each play a role in the design process. A balanced approach to important topics leads to balanced conclusions and design methodologies. The design process is affected by the experiences of junior designers with the tools that support their ability to think analytically and to emotionally regulate.

Junior designers should understand the emotional elements classified under the five categories. Strategies were developed to assist junior designers in recognising their emotional shifts, design faults and user feedback. The suggested practical techniques helped improve their motivation and self-regulation while designing. This study's methodological framework and the participants' conversations provided insight into the limitations of integrating design and emotional concepts into design education. Therefore, this study investigated strategies that can be used to teach junior designers emotional regulation throughout the creative process.

2.8.2 Emotional Configuration Model in the Design Process

While researchers have studied the evolution of emotional information during the design process, few researchers have investigated the overall development of emotional information across the entire design process from a psychological point of view, which includes identifying several separate subjects. Mattingly and Kraiger (2019) and Oliver (1993), who have researched the development of emotional information in design processes across disciplines, including psychological and design studies, have identified a number of distinct stages. Those stages include the introduction of emotional considerations into design thinking, self-awareness, self-regulation, self-motivation, empathy, innovation and EI abilities.

2.8.2.1 Self-Awareness

People do not simply transmit information through interpersonal communication, but rather they represent themselves by showing how they believe they are viewed by others, how they believe others view them, and how much importance they place on other people's opinions of them. For example, a junior designer's level of self-awareness is heavily dependent upon their perception of their performance as a designer, based on their previous education or work experience as a designer. The initial stage of the proposed model will allow junior designers to establish objectives for themselves and establish strategies to accomplish those objectives in order to enhance their performance and increase their self-esteem within the context of the design studio.

2.8.2.2 Self-Regulation

Self-regulation refers to cognitive capabilities that receive, recognise and interpret information, which people then act upon. How people interact with the world is a function of how they think about the world. As part of communicating with each other, people need to be able to observe and interpret objects, events, people and signs. People's perceptions of objects and events, therefore, directly determine how they will communicate about them, and how they communicate influences how they perceive ideas and social realities. While perception provides an excellent vehicle for communication, perception is selective and will either enhance or hinder communication. Therefore, junior designers must be provided with education on perception and communication theories related to the design process and be made aware of their perceptual limitations, i.e., what the mind does with information once it is perceived by the senses, and how to eliminate those limitations. Education on these subjects provides junior designers with the basis for decision-making and problem-solving skills. Additionally, individual and group problem-solving approaches, along with the case study approach, provide additional opportunities for developing problem-solving skills. Furthermore, junior designers must understand that interpersonal communication involves both cognition and emotion and that these two factors play a critical role in many of the decisions that are made during the design process.

2.8.2.3 Self-Motivation

In order to create high levels of self-esteem, motivation is necessary (Topçu & Leana-Taşcılar, 2018). Motivation refers to the potential of achieving objectives of a design project or task and the amount of effort applied in any particular situation. Through their natural talents, junior designers can assess their objectives and limitations and identify ways to overcome those limitations. Therefore, the performance of junior designers in their role as designers must be applicable to actual work environments, and their self-esteem must transition from the classroom to the workplace. The classroom setting affords students the opportunity to try new approaches and experiment, without fear that the outcome will significantly impact a company or their career. As such, a non-evaluative educational environment is required to enable students to discover and document their strengths and weaknesses and to establish realistic objectives, so that students can optimise their interpersonal communication skills (Al-Jedaia & Mehrez, 2020).

2.8.2.4 Empathy

The interpersonal skills and social tactics that individuals employ in their interactions with others will contribute to their ability to perform optimally (Weilenmann et al., 2018). In the classroom, students can be educated in communication skills or styles that will enable them to recognise how to analyse and apply assertive and empathic components of their performance. Additionally, students can practise establishing rapport with others and resolving conflicts. Interpersonal skills are crucial to effective interpersonal communication.

2.8.2.5 EI Design and Abilities

To effectively convey messages and describe the procedures used, designers require an ability to utilise design elements and languages. To convey themselves and to successfully complete the objectives of a project, designers must learn to utilise both verbal language and non-verbal design elements while working with others. Theories of design and emotion outline the types of presentations that can assist junior designers in achieving their objectives and the consequences of those presentations, as well as the various factors in our lives that can influence the way we utilise language (Rangel-Rodríguez et al., 2021).

Based on those theories, the following design and emotion model explains the elements that should be included in interpersonal communication training (Figure 2.3).

2.8.3 Criteria for Evaluating the Proposed Principles

To understand the effectiveness of the proposed principles for design and emotion concept configuration in the design process, it was essential to prove their ability to assist junior design students in managing emotions during the design process. Several criteria were used to evaluate the proposed principles. The participating junior design students had to progress through all five core components of the design and emotion concept configuration (self-awareness, self-regulation, internal motivation, empathy and design and emotion skills). The performance of the junior design students in managing their design process had to improve. Furthermore, the design outcomes generated through the design process, which employed the principles for design and emotion concepts, needed to be optimised. In a previous study, some research teams employed applications to evaluate the effectiveness of emotion and self-regulation training. Zotev et al. (2020) used an application to evaluate how their

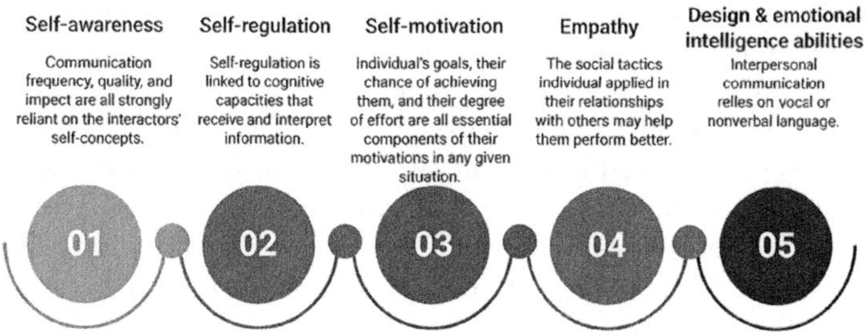

FIGURE 2.3 Five main components of management during the design process

proposed emotional self-regulation training could help patients with major depressive disorder by adopting a neuromodulator viewpoint.

In addition to quantitative methods, the participants' feedback was used to measure the effectiveness of the proposed principles. The way the junior design students critically judged, evaluated and designed visual elements, including images (Mehrabi Boshrabadi & Hosseini, 2020), was assessed, as it reflected the principles' usefulness. The research team found that design elements in visual communication can reveal meanings. In traditional teaching and learning environments, design students situate design elements in their design and art projects to contextualise their awareness of social and cultural issues and developments. New methods of teaching design history can awaken an awareness that purely chronological and/or formal accounts are often unable to achieve. Such methods can enliven debates in the classroom and increase students' engagement in a subject. Design history courses employing these ideas and techniques have the added value of imparting contextual and critical thinking.

2.8.3.1 Judgement and Evaluation Design in Visual Elements Included in Images

Mehrabi Boshrabadi and Hosseini (2020) evaluated the judgement and assessment of visual aspects, including pictures, places inside settings, critical views and models that could disclose meanings that in conventional educational contexts are not presented to design students. Extending the foundation of design history education will not reduce the traditional design concerns regarding aesthetics and style but will contextualise them within an understanding of the social and cultural setting.

Similar to biological evolutionism, human creativity is more than simply a quest for the 'latest and greatest'. Interesting questions arise when we focus on adaptive solutions to problems rather than on the criterion of value alone. For example, what is the human creativity analogue of long-term ecological sustainability? Kozbelt (2019) investigated the possibility that creative processes themselves are not historically invariant, using Dawkins and Furnham (1989) theory that evolution becomes better at evolving over time via the evolution of evolvability. In recent years, research on biological evolution, creativity, the mediating role of culture, the interaction between these aspects, and their co-evolution have begun to yield meaningful answers to fundamental human questions. These theories provide a new level of self-direction. Since the beginning of our species, evolution has shaped and constrained our ability to be creative. Creativity is ours to use in any way we see fit. The ability to be creative is unique to the human race. Almost all other animals exhibit a fixed pattern of behaviour in response to their immediate surroundings. There are examples of chimpanzees using rudimentary tools, bower birds making ornate nest structures and humpback whales and Bengalese finches altering their vocalisations. These examples, however, pale in comparison to the accomplishments of *Homo sapiens* when observing creativity across species (Kaufman, 2015). This disparity between human and animal creativity does not necessitate a fundamental separation. The gap, and perhaps the nature of creativity in general, can be better understood by considering our evolutionary history, that is, a series of evolutionary stresses and coincidences that led to the development of what we now consider to be uniquely human. As a result, our species has developed remarkable creative abilities that are, to the best of our knowledge, unmatched in the history of life on Earth.

Rodrigues et al. (2019) investigated the relationship between EI and creative thinking. Their findings indicate that good personal attitudes towards entrepreneurship and personal behavioural capital have a beneficial impact on entrepreneurial intention and that these attitudes and personal behavioural capital act as mediators between EI and entrepreneurial intention. These results indicate that spirituality has little or no impact on 'the numerous concepts this study considered. EI has a direct and beneficial impact on one's ability to be creative. The concept of emotional literacy was revealed in a study carried out by Spendlove (2007), who investigated the role of emotion in the early years of art and design education. Emotion was also the subject of his investigation of existing educational policy and practice. The research identified and conceptualised a triadic schema for theorising the position of emotion within a creative educational experience. In modern practice, the schema represents three domains: person, process and design outcome. In this context, the capacity to express one's emotional state and convey one's sentiments is called affective intelligence. Someone with a high level of emotional literacy is capable of recognising and responding appropriately to other people's emotional states. This is often viewed as a distinguishing characteristic of good partnerships.

In the United Kingdom, ensuring that students leave school emotionally literate is becoming a priority, as indicated by the development of certain programmes in both primary and secondary educational settings. These programmes include a practical manual that offers instructors ideas and interesting classroom activities that can aid students in developing emotional literacy (Spendlove, 2009). The conceptualised schema can be used within art and design education, as indicated by the national curriculum statement of significance. As suggested by the study's hypotheses, greater recognition of an emotional dimension within a triadic schema can help to develop a student's emotional capacity to engage in a creative process (person), may stimulate emotional engagement through appropriate learning contexts (process) and can facilitate emotional interfacing with outcomes (design outcome), thereby aiding the student in conceptualising the powerful interrelationships between emotion, creativity and learning.

3 Aims and Objectives

3.1 RESEARCH AIMS

Across the various fields of design, junior design students need to be trained to meet the needs of new-generation users and audiences. However, design educators are grappling with the difficulty of teaching students to apply acquired knowledge effectively and appropriately to complex issues such as emotional concerns (Goldschmidt, 2001). Educators struggle to teach students to apply cognitive skills, but designing for complex human issues additionally requires affective skills, particularly emotional changes and empathy, which present further problematic elements of design education (Walther et al., 2017). Recent work in educational psychology as it relates to learning complex cognitive thinking skills and design studio teaching methods on the use of collaborative learning experiences to communicate design and emotion concepts can potentially inform design education practices by providing empirical evidence of the reasons junior design students struggle to apply knowledge during problem-solving and suggesting instructional practices that can be employed to support effective learning in design education. This study explored the potential of implementing design and emotion concepts in the design process to teach students about the design impacts of common emotion-related problems. It employed an inquiry methodology that is commonly applied in design education in the context of industrial design, a domain that has historically focused on human factors in the built environment. The findings are situated within a larger educational context using Sweller's theory of cognitive thinking (Sweller et al., 1998), Krathwohl et al.'s (1964) proposed affective domain categorisation and Zoltowski et al.'s (2012) human-centred design. The results of the simulation experience provide valuable information to design educators across the various domains about the importance of incorporating affective learning outcomes during the early phases of design learning.

The impetus for this study arose from the collective experiences of the researchers, who, combined, have taught in design programmes for more than 30 years. Researchers across design domains have expressed frustration with the difficulty of effectively and predictably educating junior design students (Fitch, 2016; Goldschmidt, 2001; Oxman, 2001; Shadle et al., 2017; Vaughan et al., 2008). A consistent trend has been students' inability to apply knowledge learned in lecture courses that address human-centred topics such as ageing assistance, optical accessibility and universal design, as well as problem-solving in design studio projects (Borrego et al., 2010; Cross, 2001; Dorst, 2006). During problem-solving, junior design students often seem overwhelmed. They are unable to identify characteristics relevant to ageing and are incapable of creating appropriate solutions, despite having had at least one full semester of instruction on these issues. Frustrated by the apparent lack of cognitive knowledge transfer (Martinez, 2010) from lecture sessions to

DOI: 10.1201/9781003747345-3

its application in studio problem-solving and the process of creating design solutions for target populations, the researchers sought to identify how junior design students can be better educated on emotional changes and the needs of their target audience.

3.2 RESEARCH OBJECTIVES

Based on the findings of the literature review presented in the previous chapter, the objectives of this study were lightly adjusted as follows:

- To explore the approaches by which junior design students can introduce emotions into their design processes.
- To investigate which emotional elements enhance junior design students' abilities to manipulate their design process.
- To propose principles for guiding junior design students to adjust their emotional changes in the design process.
- To examine the effectiveness of principles for guiding junior design students to control their emotional changes in the design process.
- To investigate how to motivate junior design students to manipulate their design process using managed emotional elements.

4 Research Methods

4.1 RESEARCH METHODS

This study explored approaches for junior design students to introduce emotions into their design processes. To achieve this, a pilot study was conducted to collect both quantitative and qualitative data for analysis. The entire research process included two research studies, Research Study 1 and Research Study 2, each consisting of three phases.

The three phases of Research Study 1:

- Phase 1: Investigation of the emotional elements in the design process: First design project using an emotion-tracking mobile application
- Phase 2: Introduction of the proposed principles of controlling emotions in design: Lecture (scaffolding) for junior design students on accommodating emotions in the design process
- Phase 3: Questionnaire

The three phases of Research Study 2:

- Phase 1: Examination of the effectiveness of individual proposed principles: Second design project using an emotion-tracking mobile application
- Phase 2: Focus groups (14 groups, each attending a 45-minute focus group interview that includes the application of the emotion-tracking device set)
- Phase 3: Individual interviews and the reflective reviews in digital lablog

4.2 RESEARCH PARTICIPANTS

Local undergraduate students (year 2) in creative arts (design) studies in Hong Kong were invited to take part in this empirical study. The participants were unfamiliar with the concepts and knowledge of emotions (Schrage et al., 2020). These participants, who had been undergoing creative art (design) studies for two years, were an ideal sample group for the research in terms of generalising the final research results (Lum, 1997). They had learned basic concepts of design studies, such as design thinking, user experience and design outcomes. Conversely, as they had only obtained a basic understanding of the design process in a relatively short time, their knowledge was limited (Shirai et al., 2021) and had not yet shaped their beliefs about design and standard approaches used to manipulate the design process (Luckman, 1967; McLennan, 2004). The participants were unfamiliar with the concepts involved in 'design and emotion' and would gradually learn to apply these to their design processes. As such, guiding them through reflection on the design process would identify their emotional change initiatives. A total of 70 participants

DOI: 10.1201/9781003747345-4

were invited to participate in the study. In Research Study 2, they were divided into 14 groups, each consisting of five participants. Their participation in this research was voluntary and they were allowed to withdraw at any time for any reason.

4.3 RESEARCH PROCESS

The entire research process was divided into two Research Studies (1 and 2) that included a design process, a lecture to introduce the concept of design and emotion, a focus group, the application of an emotion-tracking mobile application, an interview and the use of CAPTIV L7000-Premier (an emotion-tracking device set) to collect data on various aspects (Figure 4.1).

4.3.1 Research Study 1

4.3.1.1 Phase 1: Investigation of the Emotional Elements in the Design Process

In the first stage of Research Study 1, all participants were invited to manipulate design projects individually, using an emotion-tracking application during research, material allocation and idea generation. The objective of this stage was to discover which emotional responses of junior design students enhance their judgement and evaluation (Uzun & Yıldırım, 2018) during the design process (Figure 4.2).

In Research Study 1, a mobile application, eMoods Bipolar Mood Tracker, was used to obtain a systematic review of the participants' emotional changes during their design practices. The mobile application helped participants record their opportunities and challenges in design practice. The effectiveness of this application had previously been indicated by scholars who adopted mobile applications to check for mood changes in their participants (Caldeira et al., 2017). These scholars suggested that the eMoods Bipolar Mood Tracker is a practical and user-friendly method of tracking emotional changes through reporting, graphing and logging. Therefore, this mobile application was applied to enable comprehensive research and to identify junior design students' emotional responses that enhance their decision-making in design processes (Figures 4.3–4.6).

4.3.1.2 Phase 2: Introduction of the Proposed Principles of Controlling Emotions in Design

Once the participants gained a solid understanding of their initial responses in the design process, the proposed principles were introduced. Using the scaffolding approach, a lecture on the subject was provided for the participants. After guiding the participants to an improved understanding of what types of emotions they would experience, they learned to effectively express them. Based on the five core stages of EI in the design process, the principles, aimed to improve junior design students' EI and enhance their emotional understanding, were discussed in the lecture. Exercises (as shown in Figures 4.7–4.13) were provided to improve participants' EI.

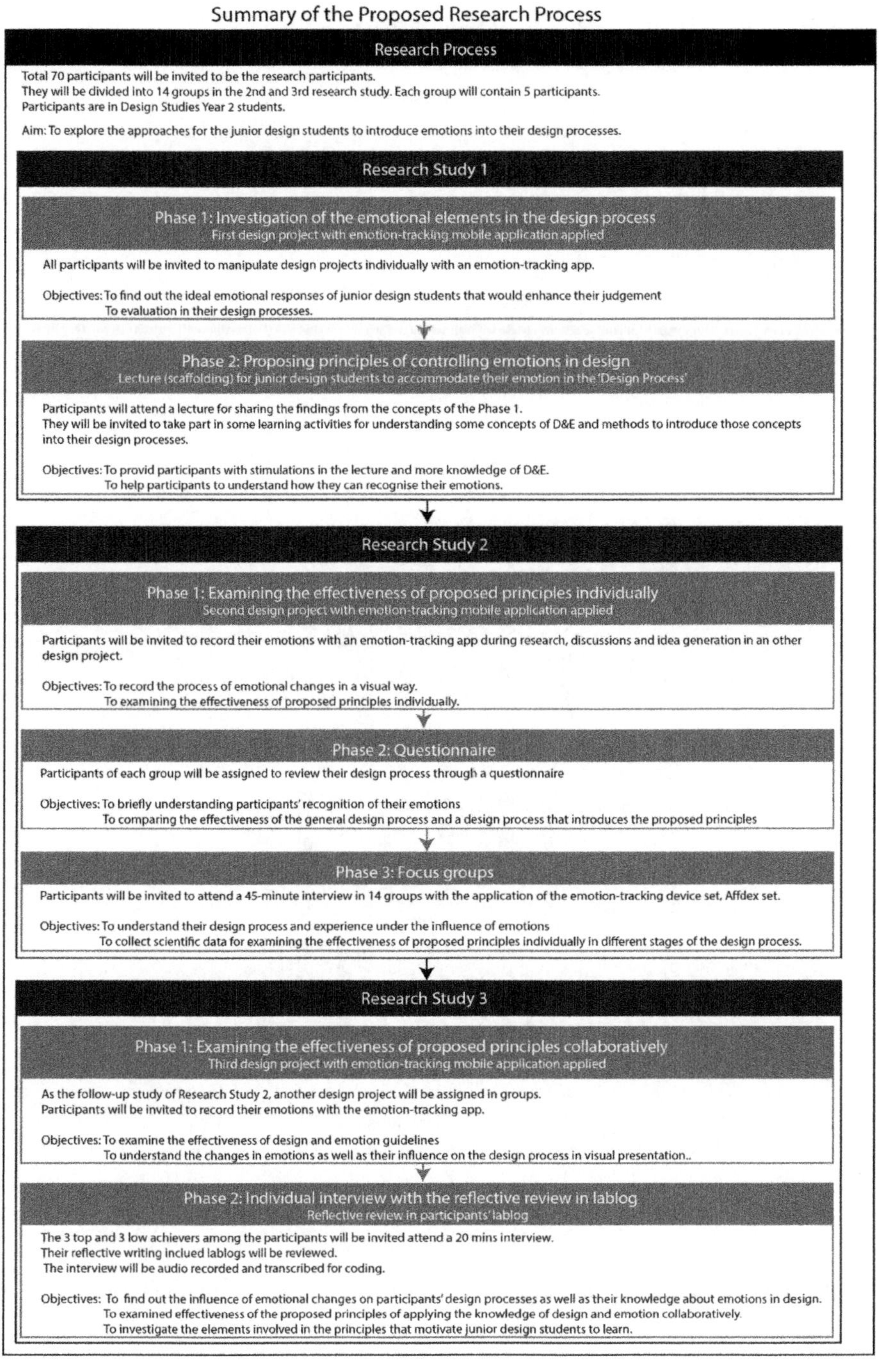

FIGURE 4.1 The research procedure

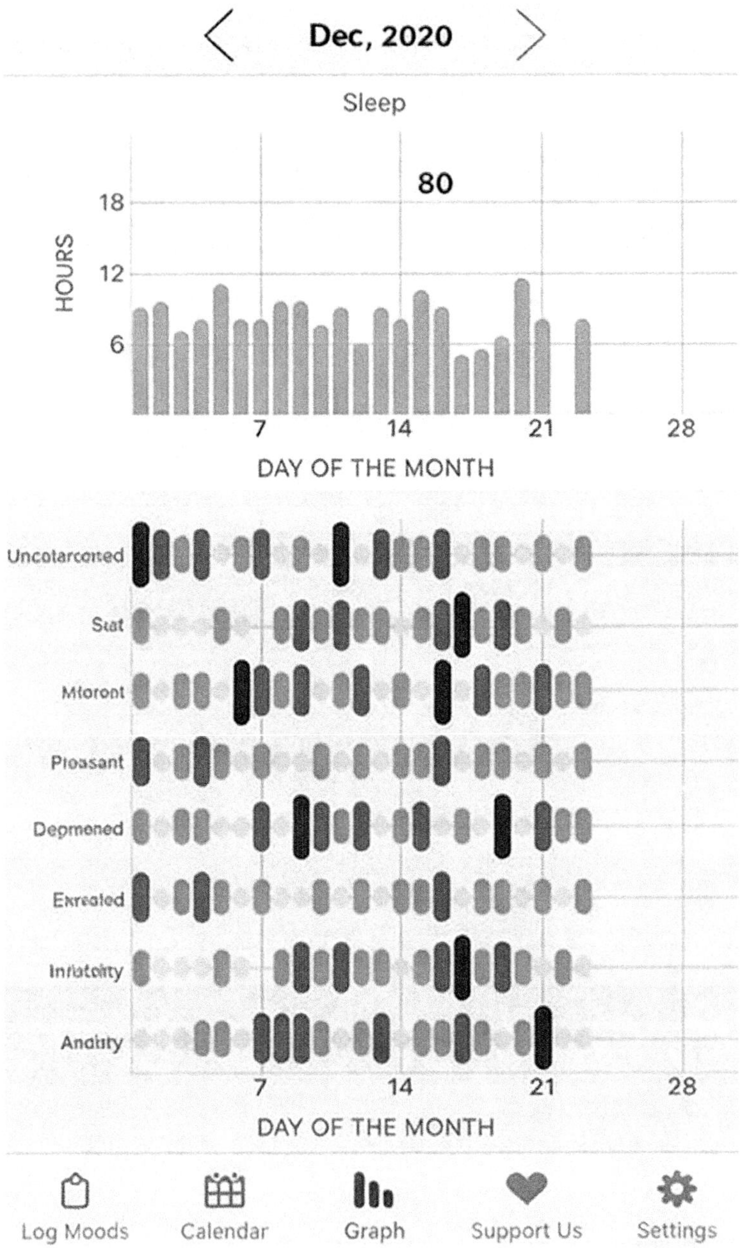

FIGURE 4.2 Screenshot of the eMoods Bipolar Mood Tracker mobile application

Please apply the app during the process of the following design exercise....

FIGURE 4.3 Instruction slide to help participants complete the assigned design exercise in Phase 1 of Research Study 1

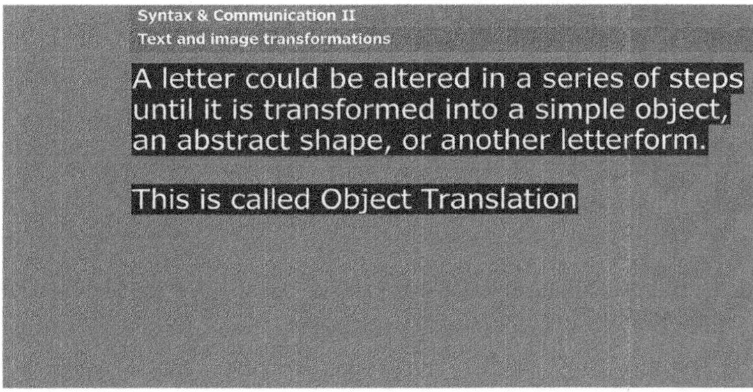

FIGURE 4.4 Instruction slide to guide participants through the assigned design exercise in Phase 1 of Research Study 1

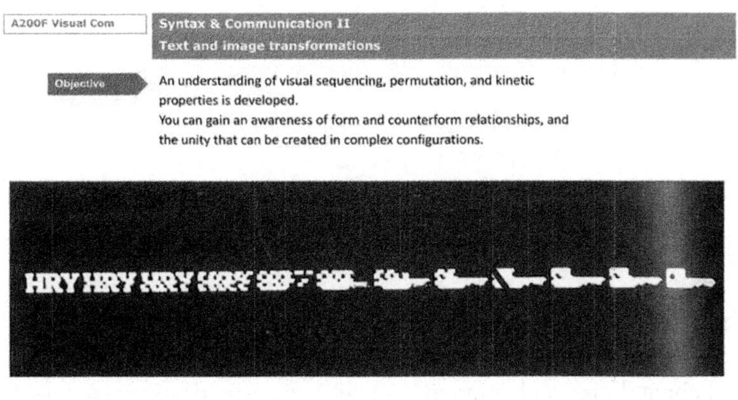

FIGURE 4.5 Instruction slide to help participants do the assigned design exercise in Phase 1 of Research Study 1

The participants obtained knowledge about ideal emotional responses that would enhance their decision-making in design processes.

Next, they were invited to take part in learning activities (scaffolding) to increase their understanding of concepts of design and emotion as well as methods to introduce those concepts into their design processes. The objectives of this stage included adding stimuli to the lecture, providing participants with additional

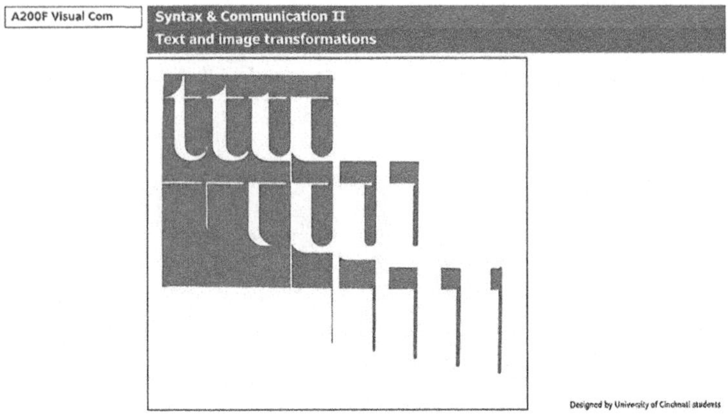

FIGURE 4.6 Instruction slide to help participants do the assigned design exercise in Phase 1 of Research Study 1

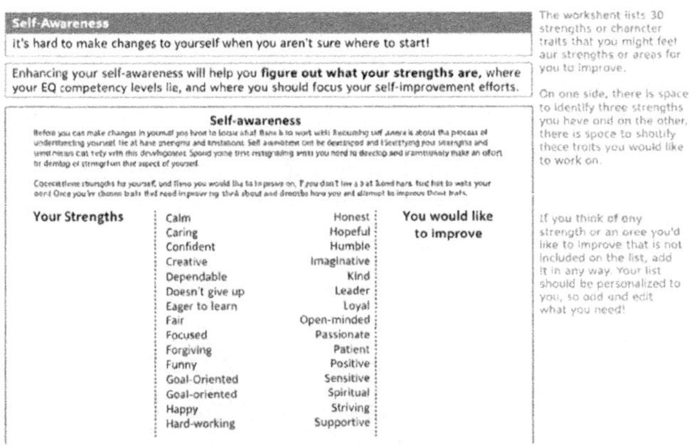

FIGURE 4.7 The exercise provided to improve participants' self-awareness

knowledge about design and emotion, and helping participants understand how they can recognise their emotions.

4.3.1.3 Phase 3: Questionnaire

In the third phase, the participants were asked to review their design process using a questionnaire designed by the researcher to collect participants' quantitative feedback. The objectives of this stage were to quickly understand participants' recognition of their emotions and to compare the effectiveness of the general design process with the one that introduced the proposed principles. The participants were asked to answer questions about the proposed principles. Their responses were kept confidential. The participants took less than 10 minutes to complete the questionnaire.

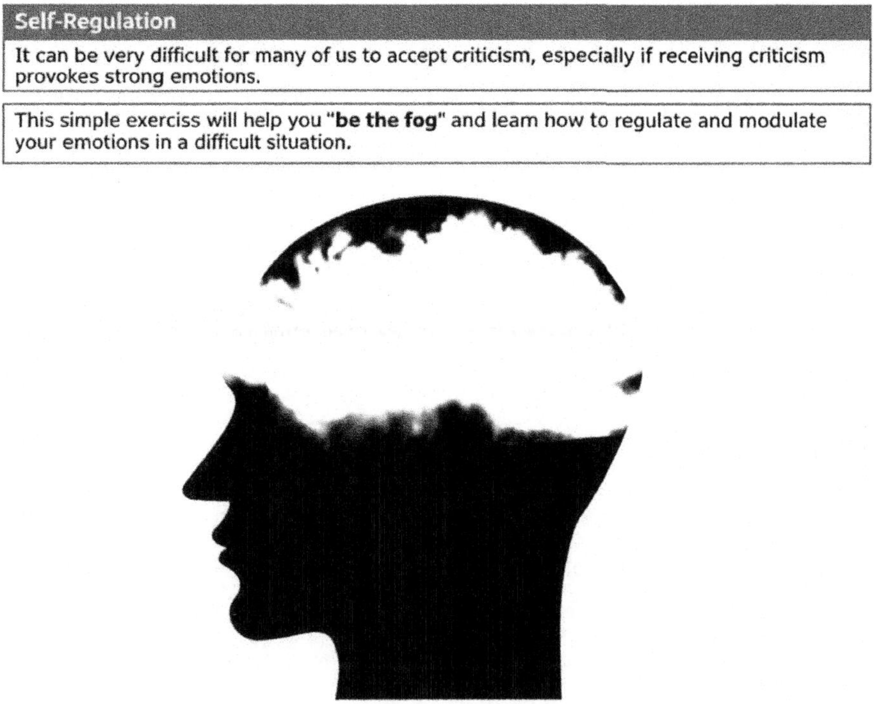

FIGURE 4.8 An exercise provided to improve participants' self-regulation (slide 1)

FIGURE 4.9 An exercise provided to improve participants' self-regulation (slide 2)

Part I: Work Effort

As it was essential to observe the participants' work efforts during Research Study 1, Part I of the questionnaire collected fundamental information regarding participation in the design projects. Participants were asked to record the 'the amount of time' they scheduled for their design process. Their input (work effort) was marked based

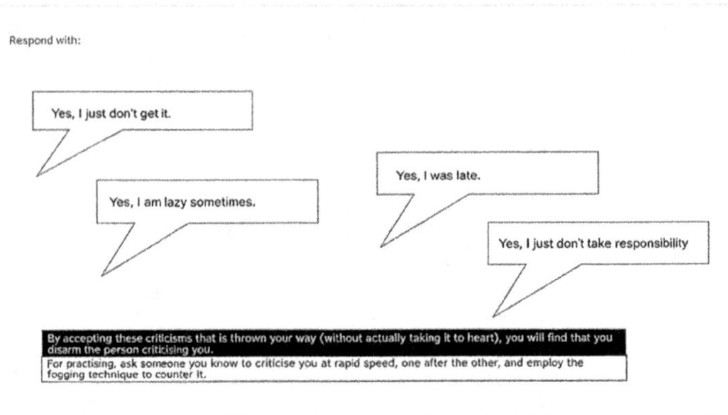

FIGURE 4.10 An exercise provided to improve participants' self-regulation (slide 3)

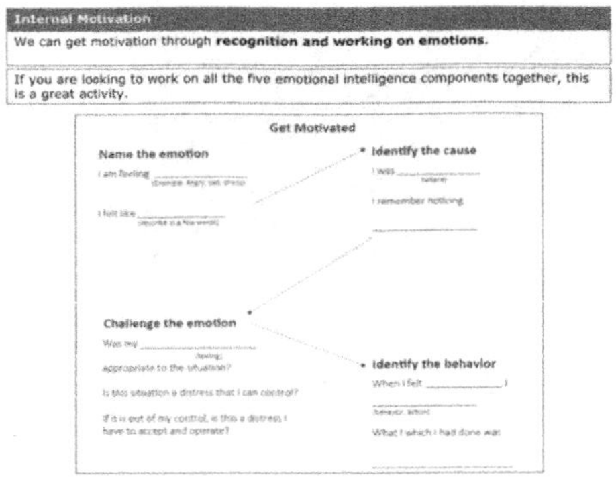

FIGURE 4.11 An exercise provided to improve participants' internal motivation

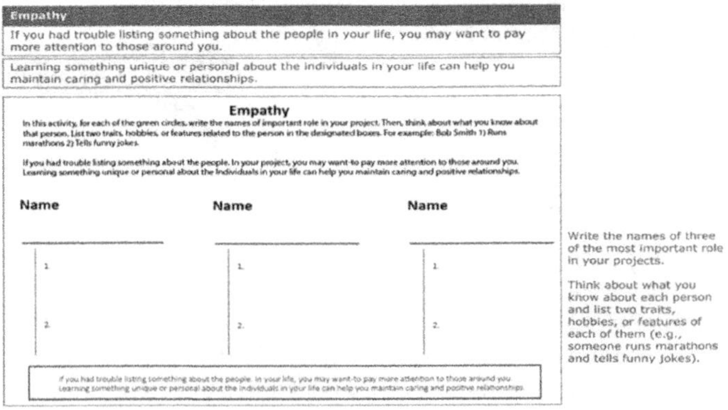

FIGURE 4.12 An exercise provided to improve participants' empathy

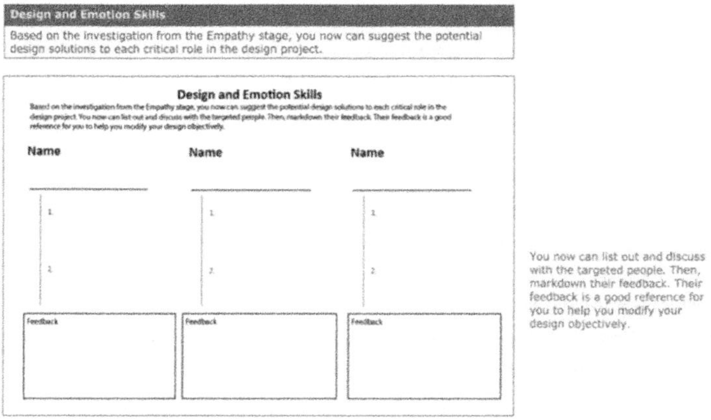

FIGURE 4.13 The exercise provided to improve participants' design and emotional skills

on their answer to the question, 'How many hours per week on average did you work on the design course (including scheduled time)?'

Part II: Emotional Levels

In Part II, the participants' emotional levels during the design process were investigated. Levels of motivation, significance, fulfilment and satisfaction related to the participants' design process were investigated to understand the significance of the process. This was reflected in the answers participants provided about the following aspects:

- Idea exploration
- Discovery and experience
- Challenges
- Belonging

Part III: Levels of Design Productivity and Complexity of the Design Process

In Part III, the levels of design productivity and complexity of the design process were addressed to obtain participants' feedback about the design process' clarity. The participants' understanding was reflected by their answers to questions on the following aspects:

- Clear objectives and organisation
- Understanding the course (Figures 4.14–4.19)

4.3.2 Research Study 2

4.3.2.1 Phase 1: Examination of the Effectiveness of Individual Proposed Principles

In the fourth stage, another design project was assigned. Participants were invited to record their emotions with an emotion-tracking app during research, discussions and

Part I. Scheduled time of your design process

Concerning the design process for the final poster design project of IDDA A200F, it is essential to first observe the participants' work efforts.

Input work effort

How many hours per week on average did you work on the design course (including scheduled time)? (hours/week)

Part II. Meaningfulness – The emotional level

Levels of motivation, significance, fulfilment and satisfaction related to design process must be investigated to understand the meaningfulness of the design process.

Regarding the design process for the final poster design project of IDDA A200F, its meaningfulness to you is reflected by the level of interest in idea exploration, exploration and personal experience, challenges and belonging.

The participants answered questions on the following aspects:
Interest in idea exploration (1), (2);
Exploration and experience (1), (2), (3), (4);
Challenges (1), (2);
Belonging (1), (2), (3), (4).

FIGURE 4.14 Questions asked to lead participants to reflect on their considerations in their design process A

idea generation. The objectives of this stage were to record the process of emotional changes visually and to examine the effectiveness of the proposed principles individually (Figures 4.20 and 4.21).

4.3.2.2 Phase 2: Focus Groups

In this stage, participants were invited to attend a 45-minute interview in 14 groups to help them understand their design process and experience as influenced by emotions, using the emotion-tracking mobile application CAPTIV L7000-Premier. In the focus groups, participants were asked to review their experience of applying the proposed emotion and design principles to their design process. The record in the emotion-tracking mobile application was reviewed. To effectively understand participants' emotional changes in the focus group discussion, a system that recognises emotional responses, the CAPTIV L7000-Premier, was applied. The application comprises a device set that can support a wide variety of aspects, such as emotional responses, eye movement and self-control behaviours (Tewari et al., 2020). The CAPTIVE-L 7000-Premier system is functionally equivalent to several

A. Level of interest in idea exploration

The level interest in idea exploration reflects the level of motivation concerning the design process.

A 1) I worked with interesting issues.

A 2) Which kind of emotion was evoked?

Anger: fury, outrage, wrath, irritability, hostility, resentment and violence.
Sadness: grief, sorrow, gloom, melancholy, despair, loneliness and depression.
Fear: anxiety, apprehension, nervousness, dread, fright and panic.
Joy: enjoyment, happiness, relief, bliss, delight, pride, thrill, joy and ecstasy.
Interest: acceptance, friendliness, trust, kindness, affection, love and devotion.
Surprise: shock, astonishment, amazement, surprise and wonder.
Disgust: contempt, disdain, scorn, aversion, distaste and revulsion.
Shame: guilt, embarrassment, chagrin, remorse, regret and contrition.

B) Exploration and personal experience

During the design process, exploration and experience influence its purposes and significance.

B 1) I explored parts of the subject on my own.

B 2) Which kind of emotion was evoked?

Anger: fury, outrage, wrath, irritability, hostility, resentment and violence.
Sadness: grief, sorrow, gloom, melancholy, despair, loneliness and depression.
Fear: anxiety, apprehension, nervousness, dread, fright and panic.
Joy: enjoyment, happiness, relief, bliss, delight, pride, thrill, joy and ecstasy.
Interest: acceptance, friendliness, trust, kindness, affection, love and devotion.
Surprise: shock, astonishment, amazement, surprise and wonder.
Disgust: contempt, disdain, scorn, aversion, distaste and revulsion.
Shame: guilt, embarrassment, chagrin, remorse, regret and contrition.

B 3) I was able to learn by trying out my own ideas.

B 4) Which kind of emotion was evoked?

Anger: fury, outrage, wrath, irritability, hostility, resentment and violence.
Sadness: grief, sorrow, gloom, melancholy, despair, loneliness and depression.
Fear: anxiety, apprehension, nervousness, dread, fright and panic.

FIGURE 4.15 Questions asked to lead participants to reflect on their considerations in their design process B

Part III. Comprehensibility – The cognitive level

The levels of productivity and complexity during the design process reflect the clarity of the design process.

Concerning the design process in the final poster design project of IDDA A200F, its comprehensibility reflects the following aspects: clear objectives and organisation and understanding of the subject.

The participants answered questions on the following aspects:
Clear objectives and organisation (1), (2), (3), (4);
Understanding of the course (1), (2), (3), (4), (5).

Clear objectives and organisation

During the design process, the clarity of the objectives and organisation is closely linked with purpose and productivity.

A 1) The intended learning outcomes helped me understand what I was expected to achieve.

A 2) Which kind of emotion was evoked?

Anger: fury, outrage, wrath, irritability, hostility, resentment and violence.
Sadness: grief, sorrow, gloom, melancholy, despair, loneliness and depression.
Fear: anxiety, apprehension, nervousness, dread, fright and panic.
Joy: enjoyment, happiness, relief, bliss, delight, pride, thrill, joy and ecstasy.
Interest: acceptance, friendliness, trust, kindness, affection, love and devotion.
Surprise: shock, astonishment, amazement, surprise and wonder.
Disgust: contempt, disdain, scorn, aversion, distaste and revulsion.
Shame: guilt, embarrassment, chagrin, remorse, regret and contrition.

A 3) I understood how the course was organised and what I was expected to do.

A 4) Which kind of emotion was evoked?

Anger: fury, outrage, wrath, irritability, hostility, resentment and violence.
Sadness: grief, sorrow, gloom, melancholy, despair, loneliness and depression.
Fear: anxiety, apprehension, nervousness, dread, fright and panic.
Joy: enjoyment, happiness, relief, bliss, delight, pride, thrill, joy and ecstasy.
Interest: acceptance, friendliness, trust, kindness, affection, love and devotion.
Surprise: shock, astonishment, amazement, surprise and wonder.
Disgust: contempt, disdain, scorn, aversion, distaste and revulsion.
Shame: guilt, embarrassment, chagrin, remorse, regret and contrition.

FIGURE 4.16 Questions asked to lead participants to reflect on their considerations in their design process C

Understanding of the subject

B 1) I understood what the teachers were talking about.

B 2) I was able to learn from concrete examples that I could relate to.

B 3) Which kind of emotion was evoked?

Anger: fury, outrage, wrath, irritability, hostility, resentment and violence.
Sadness: grief, sorrow, gloom, melancholy, despair, loneliness and depression.
Fear: anxiety, apprehension, nervousness, dread, fright and panic.
Joy: enjoyment, happiness, relief, bliss, delight, pride, thrill, joy and ecstasy.
Interest: acceptance, friendliness, trust, kindness, affection, love and devotion.
Surprise: shock, astonishment, amazement, surprise and wonder.
Disgust: contempt, disdain, scorn, aversion, distaste and revulsion.
Shame: guilt, embarrassment, chagrin, remorse, regret and contrition.

B 4) Understanding of key concepts had high priority.

B 5) Which kind of emotion was evoked?

Anger: fury, outrage, wrath, irritability, hostility, resentment and violence.
Sadness: grief, sorrow, gloom, melancholy, despair, loneliness and depression.
Fear: anxiety, apprehension, nervousness, dread, fright and panic.
Joy: enjoyment, happiness, relief, bliss, delight, pride, thrill, joy and ecstasy.
Interest: acceptance, friendliness, trust, kindness, affection, love and devotion.
Surprise: shock, astonishment, amazement, surprise and wonder.
Disgust: contempt, disdain, scorn, aversion, distaste and revulsion.
Shame: guilt, embarrassment, chagrin, remorse, regret and contrition.

FIGURE 4.17 Questions asked to lead participants to reflect on their considerations in their design process D

emotional response tracking systems, such as the CAPTIV-Neurolab System, which was adopted in studies that investigated the needs of children and adults.

These studies showed that the emotional response tracking technology is essential to empower the mental analysis algorithms of participants' psychological systems

C) Challenges

The challenge level during the idea exploration phase affects the enjoyment and fulfillment of the design process.

C 1) The course was challenging in a stimulating way.

C 2) Which kind of emotion was evoked?

Anger: fury, outrage, wrath, irritability, hostility, resentment and violence.
Sadness: grief, sorrow, gloom, melancholy, despair, loneliness and depression.
Fear: anxiety, apprehension, nervousness, dread, fright and panic.
Joy: enjoyment, happiness, relief, bliss, delight, pride, thrill, joy and ecstasy.
Interest: acceptance, friendliness, trust, kindness, affection, love and devotion.
Surprise: shock, astonishment, amazement, surprise and wonder.
Disgust: contempt, disdain, scorn, aversion, distaste and revulsion.
Shame: guilt, embarrassment, chagrin, remorse, regret and contrition.

D) Belonging

The level of belonging experienced during the course reflects the significance and satisfaction of the design process.

D 1) I experienced togetherness with others who were completing the course.

D 2) Which kind of emotion was evoked?

Anger: fury, outrage, wrath, irritability, hostility, resentment and violence.
Sadness: grief, sorrow, gloom, melancholy, despair, loneliness and depression.
Fear: anxiety, apprehension, nervousness, dread, fright and panic.
Joy: enjoyment, happiness, relief, bliss, delight, pride, thrill, joy and ecstasy.
Interest: acceptance, friendliness, trust, kindness, affection, love and devotion.
Surprise: shock, astonishment, amazement, surprise and wonder.
Disgust: contempt, disdain, scorn, aversion, distaste and revulsion.
Shame: guilt, embarrassment, chagrin, remorse, regret and contrition.

D 3) The atmosphere of the course was open and inclusive.

D 4) Which kind of emotion was evoked?

Anger: fury, outrage, wrath, irritability, hostility, resentment and violence.
Sadness: grief, sorrow, gloom, melancholy, despair, loneliness and depression.
Fear: anxiety, apprehension, nervousness, dread, fright and panic.
Joy: enjoyment, happiness, relief, bliss, delight, pride, thrill, joy and ecstasy.

FIGURE 4.18 Questions asked to lead participants to reflect on their considerations in their design process E

More questions about the experience in the design process

The following questions guided participants to consider special moments and the overall significance and productivity of the design process in the final poster design project during the IDDA A200F course.

What were some of the most interesting discoveries I made while working on this project? About the problem? About myself? About others?

a) What was my most challenging moment in the entire design process and why?

b) Which emotions were evoked during the most challenging moment in the entire design process?

Anger: fury, outrage, wrath, irritability, hostility, resentment and violence.
Sadness: grief, sorrow, gloom, melancholy, despair, loneliness and depression.
Fear: anxiety, apprehension, nervousness, dread, fright and panic.
Joy: enjoyment, happiness, relief, bliss, delight, pride, joy and ecstasy.
Interest: acceptance, friendliness, trust, kindness, affection, love and devotion.
Surprise: shock, astonishment, amazement, surprise and wonder.
Disgust: contempt, disdain, scorn, aversion, distaste and revulsion.
Shame: guilt, embarrassment, chagrin, remorse, regret and contrition.

FIGURE 4.19 Questions asked to lead participants to reflect on their considerations in their design process F

(Tracey & Hutchinson, 2019). It provides accurate facial expression and body language tracking, using concepts of emotional literacy (Noffke & Somekh, 2009). The effectiveness of the CAPTIV L7000-Premier system provided scientific data to examine the effectiveness of the proposed principles individually in different stages of the design process.

Before describing the research design further, it is essential to understand how the CAPTIV L7000-Premier system assisted this study. Neuroscience is an emerging research field. Its relationship with the science of learning is a primary research area that can contribute to the development of measurable theories to help design educators understand the feedback effects of existing teaching methods and data, thereby making it possible to establish a more effective teaching practice direction and a more effective learning model than currently in place (Benedict & Hilton, 2006). Knowledge about creativity, emotions and mirror neurons is being developed. Primarily limited to theoretical predictions, early investigations into these aspects have nonetheless made it possible to transition to a new direction and prepare a foundation for further research on improving creative potential in the classroom (Figure 4.22 and 4.23).

course coordinator & examiner

IDDA 200F Foundation of Visual Communication

Amic Ho [amicgho@hkmu.edu.hk]

Final project

Visual Communication design...

Moodboard>>

Concept Development...Thumbnail...Sketches...Layout...Design Outcome

Introduction

Seeing is a natural fact for the human being; while perception is an enabling process. Design has to deal with both of these. To see does not in any way guaranteed the ability to make understandable, functional visual statements. Intuition is not enough; it is not a mystical force in visual expression. Visual meaning as conveyed by manipulation of visual elements are fundamentals of visual communication design.

This project will provide students with a big picture of the roles & functions of visual communication and how does it differ from other design disciplines. Students can understand how visual elements, colours, typefaces, compositions could interact to form communication pieces of a poster that create messages and hence to develop the visual design as a kind of strong and powerful way of communication.

Objectives

- To apply communication principles with visual elements to deliver messages.
- To explore the visual syntax and communication in compositional elements, media and information.
- To learn about the design process of visual communication design steps by steps and the close relationship of each other.
- To realise how to communicate with visuals, images and graphic elements.

samples of poster design retrieved from

FIGURE 4.20 Project brief for the second design project using an emotion-tracking mobile application in Research Study 2 Phase 1 A

Part I
Mataphor Concept Development

You are going to design an A2 poster.

Base on the concept and mood board that you had designed and the 3 concepts that you had developed, review and select one of them that could use to communicate effectively and further develop to the topic/message of

Part II
Montage Concept Visualisation

Thumbnails

- Show 30 thumbnails of layout concepts for your poster.
- Prepare any visual reference that required to explain clearly your concepts such as illustration styles or photographic styles to support your thumbnails

>>> **Deliverable: 30 thumbnails**

Layout Sketches

- Develop layout sketches of the chosen thumbnail(s).
- It should be the variations of your chosen thumbnail(s) such as colour variations, font variation or graphic styles. The layout sketches should be hand-drawn

>>> **Deliverable: Layout sketches**

Part III
Masterpiece Concept Execution

Full-size poster

- Develop the full-size A2 poster.
- It can be A4 print-outs joined together or simply your big cut & paste layout.

Design refinement

- By this time, the poster design should have been finished and finalised.
- It is done for checking readability and legibility, the proportion in the spatial relationship of visual elements and the overall impact of the poster.

Exhibit your finalised poster

- Execute the poster design by bringing it to the digital output. (Please consult and confirm with your supervisor the way to execute to the poster due to COVID-19 pandemic)
- Display and present your poster in the final presentation.

>>> **Deliverable: Poster design**

Design Requirements

1. The whole development process including the thumbnail and layout sketches should be included and present with Google slides with clear structure and hierarchy.
2. The presentation slides and proposal should be well-designed and able to reflect to the theme or philosophy of your poster.
3. There is always a reason in each design initiatives and be prepare to explain it. Fun and interesting is not the only reason.
4. Proper file management and naming system for all submissions in the format of: Student Full Name_Student no._IDDA A200F project name

FIGURE 4.21 Project brief for the second design project using an emotion-tracking mobile application in Research Study 2 Phase 1 B

Measuring Emotional Responses through Psychological Devices

Once they better understand the categorisation of emotions, it will be possible for designers to identify their emotional changes to some degree. However, most investigations still require a narrow focus on simple emotions with a single label, and they ignore changes because there is a shortage of trustworthy datasets. An accurate

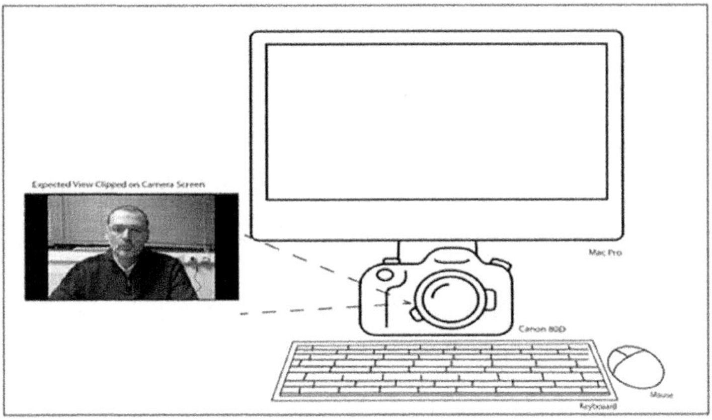

FIGURE 4.22 Equipment setting for each student in the focus group

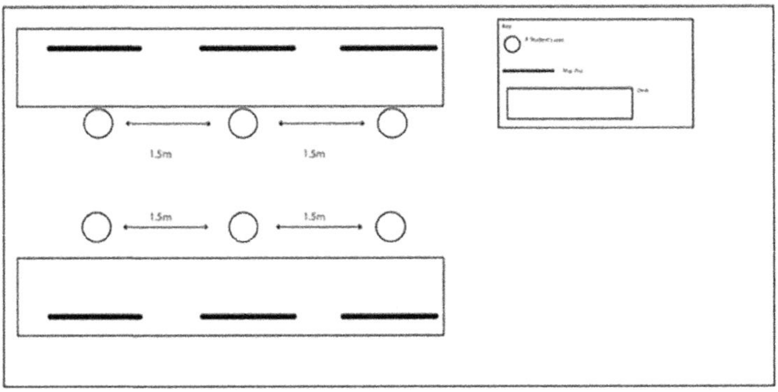

FIGURE 4.23 Seating plan for the focus group

database must be developed to obtain an equivalent reference to effectively recognise various emotional changes. Measurements are needed to illustrate basic emotional responses and changes.

Electrodermal Activity Measuring Tools

The study of emotion has standardised measurements, induction protocols, data processing tools, sub-disciplines and fundamental theoretical issues. Various emotion induction methods, including electrodermal activity measuring tools, have been developed to elicit and manipulate emotional reactions. These tools detect changes in the physical body based on minute reactions. The connection between body conductance and emotion was discovered in the 19th century (Ho & Siu, 2012). It can be difficult for people to disconnect emotional reactions from the autonomic nervous system (which is physically linked to the skin's surface). This connection is frequently used in psychological investigations and is one of the simplest and most reliable psychophysiological measurements used to evaluate physiological activation in emotional induction.

Electrodermal activity measuring tools evaluate electrical activity occurring on or emanating from the facial muscles and the skin's surface. These tools have been used in psychological research for over a century and remain some of the simplest and most accurate psychophysiological tools used to measure physiological engagement in emotional induction (Williams et al., 2022). A connection between skin conductance and emotion was discovered and soon after employed in a study (Pineau et al., 2022). Several terms were used to characterise the skin's active and passive electrical properties, including 'skin conductance response', 'sympathetic skin response' and 'skin conductance level' (Tronstad et al., 2022). Recently, these terms were standardised as 'electrodermal activity measuring tools' (Bordone et al, 2019). The skin's reaction to small electric currents can be measured and identify nervous behaviour. The autonomic nervous system may be activated if the current flows easily between two sites on the skin's surface. Perspiration influences skin conductivity. Due to parasympathetic autonomic nervous system activation, agitated or emotional people perspire more than those not experiencing those feelings, thus increasing skin moisture. This improves electric current flow, which reflects emotional involvement.

To reveal the efficiency of using electrodermal activity to measure emotional arousal, it is essential to understand the function of applying this system. Under the umbrella term 'electrodermal activity measuring tools', the sweat circuit model of skin conductance is the most popular application (Tronstad et al., 2022). This model posits that phasic skin conductance fluctuations occur when epidermal sweat ducts are full. Skin conductance returns to tonic values when fluid is deposited on or fully absorbed by the sweat glands. As the sweat glands fill with perspiration, the sweat ducts will serve as 'variable resistors'. The number of sweat glands activated and the volume of sweat supplied to the ducts will determine the conductance change (Turner, 2021). Brain-controlled sympathetic autonomic nervous system activation of the sweat glands is a basic survival strategy. Although an electrical shift alone may not identify precise emotions, certain patterns coincide with fundamental emotions, such as pleasure, rage or sorrow (Kreibig & Gendolla, 2014). The degree of skin conductance will decrease or increase based on such emotions. The link between sweat gland activity, the autonomic nervous system and electrical conductance are excellent indicators of emotional arousal. As an electrodermal activity measuring tool, the facial expression tracker can effectively track emotional arousal.

Analysis of Facial Expressions

An important non-verbal cue used in human communication to anticipate intention is facial expression. According to Mehrabian's (2017) research, 55% of signals relating to sentiments and attitudes are conveyed via facial expression, the remainder are expressed verbally and 40% are made known by behaviour. According to Mehrabian, facial expression is crucial to information transmission. Automatic facial expression recognition has been closely investigated in recent years due to the rapid advancement of artificial intelligence. Psychology, computer technology and pattern recognition have highlighted the research on facial expression recognition.

Human–computer interaction (Chatterjee et al., 2019; Saxena et al., 2020), virtual reality (Chen et al., 2015; Li, 2022), augmented reality (Assari & Rahmati, 2011; Chen et al., 2015), sophisticated driver assistance systems (Assari & Rahmati, 2011), education and entertainment (Lankes et al., 2008) are just a few of the many fields in which facial expression recognition has extensive applicability. For technologies involving electrodermal activity, facial expression analysis is a measurement that has gained considerable interest for its many potential uses in human–computer interaction. Using the link between facial expression and the autonomic nervous system, a face tracker was designed as a reflective device to measure changes in facial expression.

Facial expression is a basic and frequently used method to express emotion and intention (Darwin & Prodger, 1998; Li & Deng, 2020; Tian et al., 2001). Numerous studies have been conducted to analyse automated facial expressions based on their importance in social robotics, medical treatment, driver fatigue monitoring and other human–computer interface systems. Many facial expression detection approaches have been investigated in machine learning and computer vision to extract expression information from images of faces. Based on cross-cultural research (Ekman, 1994) that shows that people experience certain fundamental emotions the same way, regardless of culture, Ekman and Friesen (1971) denoted six basic emotions, namely anger, contempt, fear, pleasure, sorrow and surprise, as early as the 20th century (Ambrose et al., 2019). According to recent findings from modern neuroscience and psychology studies, the paradigm of the six basic emotions is culturally specific rather than universal. The GSR eFace Coding database is the most often used lab-controlled database that contains sample video clips for studies on emotional tracking in general and specific, prepared clips for certain subjects. The segments show a shift in facial expression from neutral to peak emotion, and they vary in duration from 10 to 60 frames. This system involves the typical emotion categories proposed by Ekman and Friesen (1971), including anger, sadness, surprise and happiness (Ambrose et al., 2019), and employs the most common data extraction method for static-based techniques. However, the algorithms evaluated by the GSR eFace Coding database are incoherent, as they lack particular validation and datasets. The sequence of facial expressions is recorded. To analyse slight changes in facial expression in the design process, a more mature coding system with a sufficient database is needed.

Within the applied ergonomics technologies system, the CAPTIV L7000-Premier facial expression tracker is used to monitor 49 marked points on the face and identify characteristics using a regression model. The 49 points serve as markers that are identified using a regression model. The translation and scaling differences between the original and actual marked positions are then calculated. The primary concept behind feature tracking is the supervised descent method, which is used for marker estimation in each frame. The model outputs the coordinates for fiducial points after training using 66 landmarks. The computed and averaged distances and variations between the eyebrows, eyes and lips to the reference point represent emotional responses. After the head posture is adjusted, the facial points are extracted. Each subject's neutral facial expression and the mean expression of all the subjects are determined to have an affinal transformation in common, and this transformation is used to register the tracked points in the sequences. The tracked points are created by averaging the positions of the markers on the inner corners of the nose and eyes (Soleymani et al., 2015). Slight changes in facial expression can thus be detected. The CAPTIV L7000-Premier application was therefore used to conduct the following research stages.

4.3.2.3 Phase 3: Individual Interviews and the Reflective Reviews in Lablog

In this stage, the three highest and the three lowest achievers among the research participants were invited to attend a 20-minute interview and lablog review to gain a better understanding of their emotional changes and the influence of emotions on their design process. This phase was audio-recorded and transcribed to allow the researcher to gain a robust understanding of the construct. The coding was analysed both qualitatively and quantitatively, and coding schemes were developed.

The digital lablog included a complete record of the learning process and served as a tool with which to try new things, a reflective journal and a thinking tool. Each participant's digital lablog was a detailed account of the order and duration of all the aspects they focused on during the design process and enabled them to use their time effectively. Additionally, after reflection, the lablog helped to ensure adequate time management for future projects. As a tool with which to try new things, it served as a digital notebook for the recording of dates and places, people encountered, observations of phenomena, new thoughts and feelings, experiences and knowledge, and journeys of discovery. As a reflective journal, it provided an invaluable database for revising and improving participants' thinking and problem-solving processes (Murty & Purcell, 2007). The lablog provided a record of thoughts, impressions, opinions, action strategies, perceived outcomes and results (Tracy & Robins, 2008). It was essential for highlighting possibilities and expanding opportunities. As a thinking tool, revisiting and continuously examining the participants' creative processes (Uwajeh, et al., 2019). meant that the lablog provided a platform for generating ideas and a portable, convenient space in which to record participants' mind maps, brainstorming lists, answers to questions, concept development and design analysis.

For this study, the digital lablog was the best option to gain conditional knowledge, as it recorded all the information about the learning process. It helped participants learn the reasons for and approaches to applying knowledge declaratively and procedurally. It helped them decide on resource allocation strategically. Thus, their strategies or design solutions were strengthened and more effective than before. Adopting the digital lablog encouraged participants to evaluate how the proposed principles assisted them in conducting the design process using metacognition. Indeed, the concept of metacognition supported participants' insightful feedback to the design research field (Yim & Byon, 2018). The process of planning, applying and evaluating (involving all thinking activities and problem-solving and evaluation procedures) using the digital lablog enhanced participants' knowledge and awareness and regulated their progress through evaluation.

By preparing the presentation on metacognition, insights into leading metacognitive strategies in learning were obtained. In Phase 3, questions about the participants' experiences in the design process were discussed. The above questions helped participants identify significant design milestones, assess their productivity in completing a design task, and evaluate their personal (individual) and professional (collaborative) experience with respect to the design challenge as a whole. Firstly, they provided an account of what they thought were the most interesting discoveries they had encountered throughout the project, including their developing perception of the problem they were attempting to solve, as well as their individual and collective experience of working through the challenges they faced. Secondly, they identified the most difficult point in the entire design process and stated why they found that particular period so difficult. Thirdly, they assessed the degree to which their emotions changed during the time when they experienced the greatest difficulty in the design process. Finally, they were asked to provide their opinions regarding how effectively their proposed design solutions addressed real-world contexts and problems.

To conclude, the digital lablog was the best tool to track the participants' emotional changes and find answers to all possible questions related to this research. Additionally, the knowledge obtained about understanding design was employed for different tasks. Participants learned effective approaches to communicating with their peers, who improved their interpersonal skills as well. This learning experience introduced the application of metacognitive control and self-regulatory processes to the participants' design learning processes. They applied cognitive processes to control and regulate their emotions while learning and, as a result, were able to improve planning, monitoring and evaluation processes. The knowledge and experience obtained from this learning activity included awareness of and reflection on the investigative methods used to obtain the actual content, that is, deep knowledge versus partial knowledge, which led to improved design outcomes. The objectives of this stage were to establish the influence of emotional changes on participants' design processes and their knowledge about emotion in design. The effectiveness of implementing the proposed principles, that is, applying knowledge of design and emotion collaboratively, was examined. The principles' elements that motivate junior design students to learn were also investigated.

5 Research Findings

5.1 OVERVIEW

Students must be prepared to fulfil the demands of learning how to incorporate emotion into design in studies pertaining to the many design professions. However, design educators face challenges in helping students apply newly gained information to complicated emotional issues in an acceptable and effective manner. Not only do educators struggle to teach students to apply cognitive skills, designing for complex human issues also requires affective skills, particularly empathy, which is another problematic element of design education (Walther et al., 2017). Educational Psychology's recent research into obtaining complex cognitive skills, and medical training using Immersive Learning to train doctors to be empathetic, gives design education a wealth of insight. They also provide empirical data as to why students have difficulty applying what they know to problems and provide instructional techniques for improving the effectiveness of learning in design; therefore, the purpose of the current study is to explore how to help young design students manage their own design process by managing their emotions effectively. The findings are situated in a larger educational context using Sweller's cognitive load theory (Sweller et al., 1998), Krathwohl's affective domain taxonomy) and Zoltowski's human-centred design (Zoltowski et al., 2012). The results of the simulation experience provide design educators across domains with valuable information about the importance of incorporating affective learning outcomes during the early phases of design learning (Zanganeh & Hariri, 2018).

Students have shown a common area of weakness, which is translating the design-based understanding of human-centred topics – such as ageing, light and the ageing eye, accessibility, universal design and problem-solving – into their studio project designs, despite having completed at least one full semester of instruction regarding each subject. Many students experience anxiety during problem-solving, and lack sufficient knowledge regarding the physical aspects of ageing and potential solutions for those challenges. Frustrated by the apparent lack of cognitive knowledge transfer (Martinez, 2010) from lecture to studio problem-solving and the process of creating design solutions for older populations, the researchers sought to identify how novice designers can be accurately and adequately educated about the visual and physiological problems encountered by older adults with eye diseases and low-vision problems (Zhou et al., 2021). We are not alone in this quest. Researchers across design domains have expressed frustration with the difficulty of effectively and predictably educating young designers for decades (Borrego et al., 2010; Cross, 2001; Dorst, 2006; Fitch, 2016; Goldschmidt, 2001; Oxman, 2001; Shadle et al., 2017; Vaughan et al., 2008).

DOI: 10.1201/9781003747345-5

5.2 RESEARCH STUDY 1

5.2.1 Phase 1: Investigation of the Emotional Elements in the Design Process

After collecting the participants' feedback for Research Study 1, the data were gathered and presented using the following charts. Figure 5.1 shows that the participants' work effort input in Research Study 1 is similar. In other words, the participants all experienced the workload in Research Study 1 in a similar manner.

5.2.1.1 Research Findings from Research Study 1 (Application Tracking)

Participants documented the experience of their emotion(s) in real time during Research Study 1, by completing the design task (which was provided to them) and recording their emotions utilising the eMoods Bipolar Mood Tracker. Participants identified a total of eight emotions including depression, elevated mood, interest, pleasure, anxiety, sadness, irritability and mood. Three participants identified anxiety, four identified irritability, three identified depression, seven identified an elevated mood, two identified pleasure and seven identified interest. Seven participants did not identify sadness. Five participants stated that they were experiencing an emotion which was unidentified (Figures 5.2, 5.3 and 5.4).

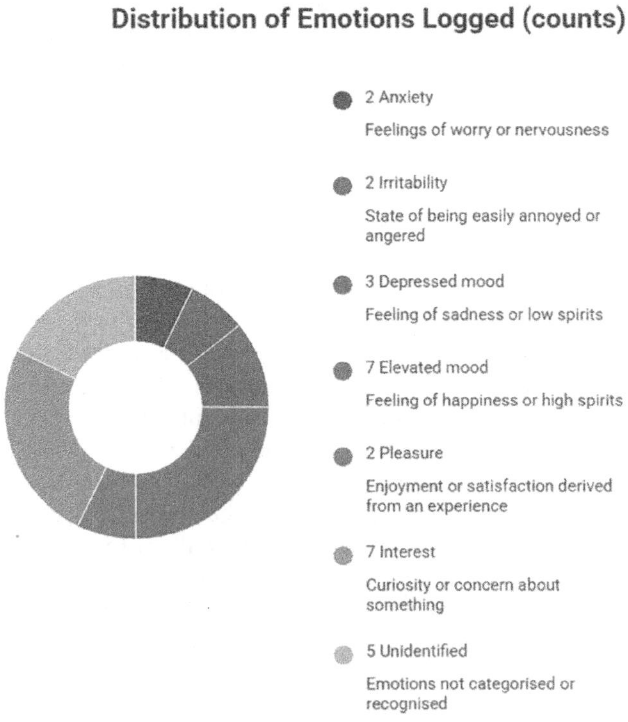

FIGURE 5.1 Recorded Work Effort 1

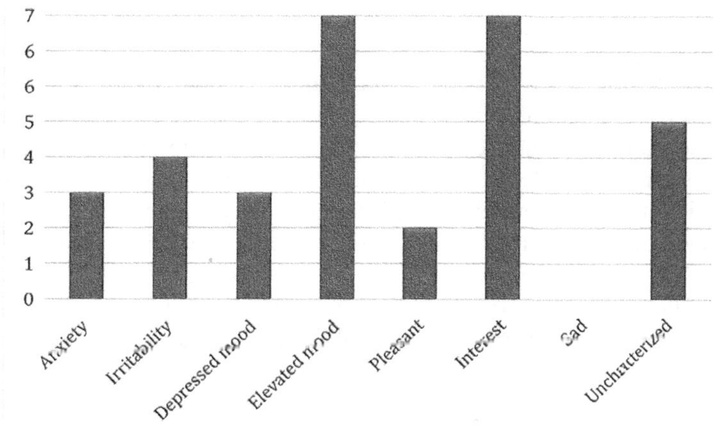

FIGURE 5.2 The emotions identified using the eMoods Bipolar Mood Tracker in Research Study 1

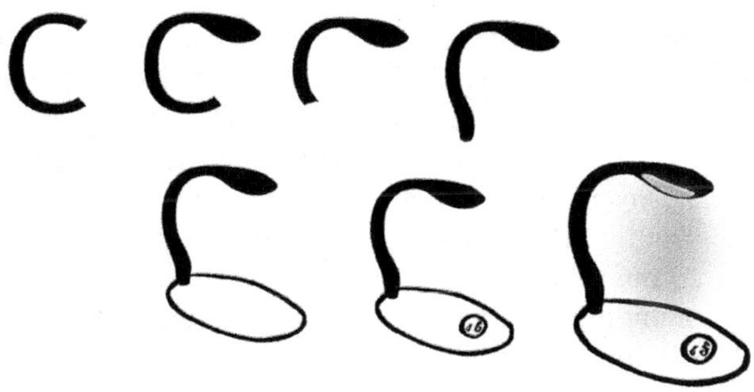

FIGURE 5.3 An example of a participant's work (designed by Participant 1)

FIGURE 5.4 An example of a participant's work (designed by Participant 2)

5.2.2 Phase 2: Introduction of the Proposed Principles of Controlling Emotions in Design

During Stage 2 of the research study, after attending an introductory presentation (during which participants learned about the results from Phase 1 of Research Study 1), participants attended a second stage where they received a 'scaffolding' presentation on how to include their feelings within the design process. The presentation provided participants with knowledge regarding what would be considered ideal emotional responses to use to make better decisions while designing. After attending the presentation, participants completed a series of learning exercises to further develop their knowledge of both Design and Emotion, and also to learn how to apply the principles of both subjects to the participants' own design process. Participants' goals for this stage of the research study were to keep the participants engaged throughout the lecture, increase their knowledge base of Design and Emotion, and assist participants to recognise their emotional state. As a result of completing the proposed methods for managing emotions in design during Stage 2 of the research study, participants successfully used the learned methods to manage their emotional responses to better control their reactions.

After summarising theories from existing psychological research and design and emotion studies (Bilevicius et al., 2018; Southward et al., 2021; Wagstaff et al., 2013), the principles of design and emotion configuration during the design process were identified as self-awareness, self-regulation, internal motivation, empathy and design and emotion skills. *Self-awareness* (Immordino-Yang, 2011) refers to understanding self-emotional changes and their effect on others. *Self-regulation* (McClelland et al., 2010) refers to controlling disruptive impulses, thinking before taking action and considering the effect of expressions and self-emotional changes on others. *Internal motivation* (Reeve, 2018) refers to working for internal reasons that go beyond profit and status. *Empathy* (Zaki, 2020) refers to understanding end-users' and the audiences' emotions and responses. *Design and emotion skills* (Van Gorp & Adams, 2012) refers to controlling relationships as well as building up emotional attachment with the audience through design implementation. The five elements above illustrate how the ideas of combining design and emotion will be incorporated into the overall design process. Building upon this foundational thinking, Phase 2 of the project began without collecting any data. The main goal of this phase was to motivate participants and increase their knowledge of design and emotion so that they could better understand and control their own emotions.

5.2.2.1 Task 1: The 'Personal Attribute List' Task to Improve Participant Self-Awareness

Participants created lists of three positive characteristics and three areas where they need to improve in terms of their self-motivation while working through the design process. After identifying the positive and negative characteristics, participants were asked to identify ways to improve the positive aspects and decrease or eliminate the negative aspects.

5.2.2.2 Task 2: The 'Put on a New Pair of Shoes' Task to Improve Self-Regulation (Perspective Taking)

As part of an exercise designed to enhance participants' ability to take the perspective of others, participants were assigned in pairs to discuss a predetermined topic for five minutes. Each participant was then asked to write down as much detail as possible about what they thought their partner was saying, including specific examples. After writing the descriptions of each other's thoughts, participants exchanged their writings to see if the written description accurately reflected the other participant's thoughts and to begin to establish a greater understanding of the other participant's viewpoint.

5.2.2.3 Task 3: The 'Get Motivated' Task to Improve Self-Motivation

This task required participants to research and present design solutions to the client for all key project functions that may have some design limitations. This task was designed to help participants develop innovative methods to overcome problems and improve their problem-solving skills.

5.2.2.4 Task 4: The 'Adjusting Your Response' Task to Improve Empathy

Prior to developing any design goals, participants identified which emotions were triggered and what goals were being pursued when responding to their clients' needs. Then, based on those emotional triggers and goals, participants developed and documented their responses to the client needs. Based on those documented responses, participants and their partners discussed their responses, provided feedback and developed refined design goals. To ensure that participants understood one another, partners restated each other's statements verbatim to confirm that the messages communicated were accurately understood.

5.2.2.5 Task 5: The 'Just Quoting' Task to Develop Design and Emotional Intelligence Skills

In this task, participants expressed their viewpoints and received paraphrased feedback from their partners. As a result of this process, participants gained a greater understanding of each other's views and the ability to effectively communicate with each other (Tables 5.1and 5.2).

5.2.3 Phase 3: Questionnaire

In the third stage, the participants in each group were asked to review their design process through a questionnaire designed by the researcher to collect participants' quantitative feedback. The participants learned strategies to manage emotions in the design process. The objective of this stage was to quickly gain an understanding of participants' recognition of their emotions and compare the effectiveness of the general design process and a design process that introduced the proposed principles. The participants were asked to answer questions about the design principles. The participants took less than 10 minutes to complete the questionnaire. Their participation in this research was voluntary, and they were allowed to withdraw at any time for any reason. Their responses will be kept confidential.

5.2.3.1 Part I. Scheduled Time of Your Design Process

To assess the design process of the final poster design project of IDDA A200F, it was essential to first review the participants' work effort.

TABLE 5.1
Randomly extracted feedback from the 'Self-awareness' worksheet

Participant	Identified strengths of the participants	Identified aspects that the participants would like to improve
A	Eager to learn, passionate	Confident, focused, open-minded
B	Fair, goal-oriented, leader	Patient, supportive
C	Open-minded, creative, funny	Kind, caring, humble
D	Humble, imaginative	Hopeful, eager to learn

TABLE 5.2
Randomly extracted feedback from the 'Get Motivated' worksheet completed by the participants

Participant	Identified emotions	Identified cause	Identified behaviour to control the emotion
A	Stress (Qin et al., 2020)	Campus working environment	Listening to K-pop
B	Anger (Briñol et al., 2018)	'Free-rider' (i.e., a person or company that gets an advantage without paying for it or earning it) teammates	Playing a mobile game
C	Excitement (Tracey & Hutchinson, 2019)	My dance society activities	Referring to assignment notes
D	Anxiety (Reinholt et al., 2022)	Failed a test	I don't know yet … maybe sleeping is a good approach

Input Work Effort

How many hours per week on average did you work on the design course (including scheduled time)? (hours/week) (Figure 5.5).

5.2.3.2 Part II: Meaningfulness – The Emotional Level

Meaningfulness

The levels of motivation, significance, fulfilment and satisfaction of the design process needed to be investigated to understand the meaningfulness of the design process. In the design process for the final poster design project of IDDA A200F, Foundation of Visual Communication BFA (Hons), the Imaging Design and Digital

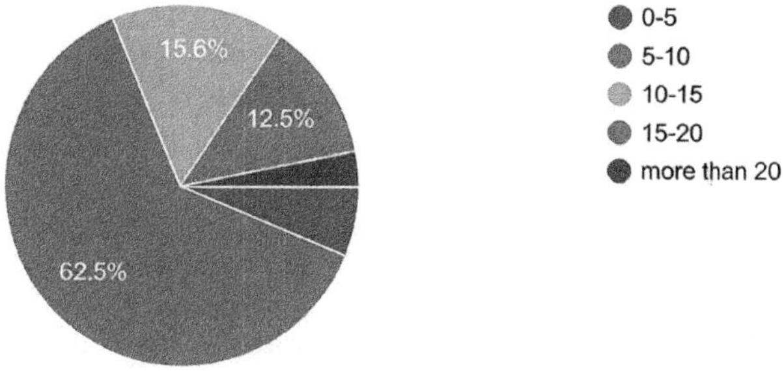

FIGURE 5.5 Participants' input in average hours per week

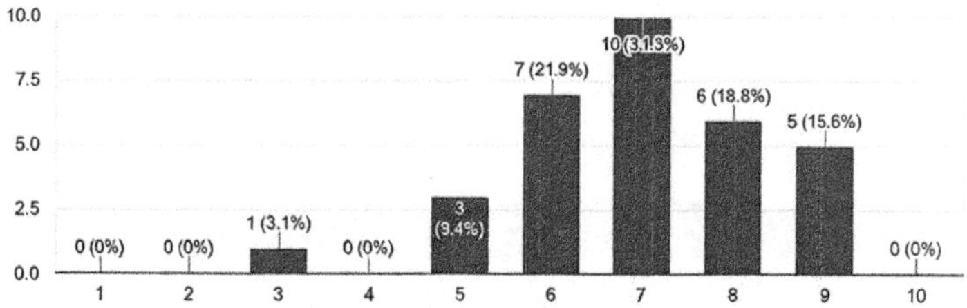

FIGURE 5.6 Participants' level of interest in idea exploration, reflecting their motivation during the design process

Art programme, its meaningfulness to participants was reflected by several aspects, namely the level of interest in idea exploration, the level of exploration and personal experience, the challenge level and the level of belonging. Accordingly, the participants were asked to answer questions about these aspects.

A. The level of interest and idea exploration

The level of interest in idea exploration reflects the participants' motivation during the design process.

A1) I worked with interesting issues (Figure 5.6).

Participants' motivation to engage in the exploration of their ideas is represented by the degree of interest that was shown during the design process. Interest levels for the issues and ideas that were examined ranged from 0 (no interest) to 9 (high interest). Approximately 66% of the participants indicated an interest level of seven or greater when examining the issues and ideas they had considered, and therefore had a strong motivation to examine them. A slightly lower number of the participants (approximately 32%) indicated an interest level ranging from 5–6, which indicated a moderate interest level and motivation; while approximately 3% indicated a low interest level (level 3).

A2) Which kind of emotion was evoked?

- aroused, astonished, stimulated, surprised, active

- enthusiastic, elated, excited, euphoric, lively, peppy
- happy, delighted, glad, cheerful, warm-hearted, pleased
- relaxed, content, at rest, calm, serene, at ease
- quiet, tranquil, still, idle, passive
- dull, tired, drowsy, sluggish, bored, droopy
- unhappy, miserable, sad, grouchy, gloomy, blue
- distressed, annoyed, fearful, nervous, jittery, anxious (Figure 5.7)

Sixty per cent of the participants expressed a positive emotional connection during their interaction with the design. The positive emotions expressed by the participants included acceptance, friendliness, trust, kindness, affection, love and devotion. Conversely, none of the participants expressed an emotion that is associated

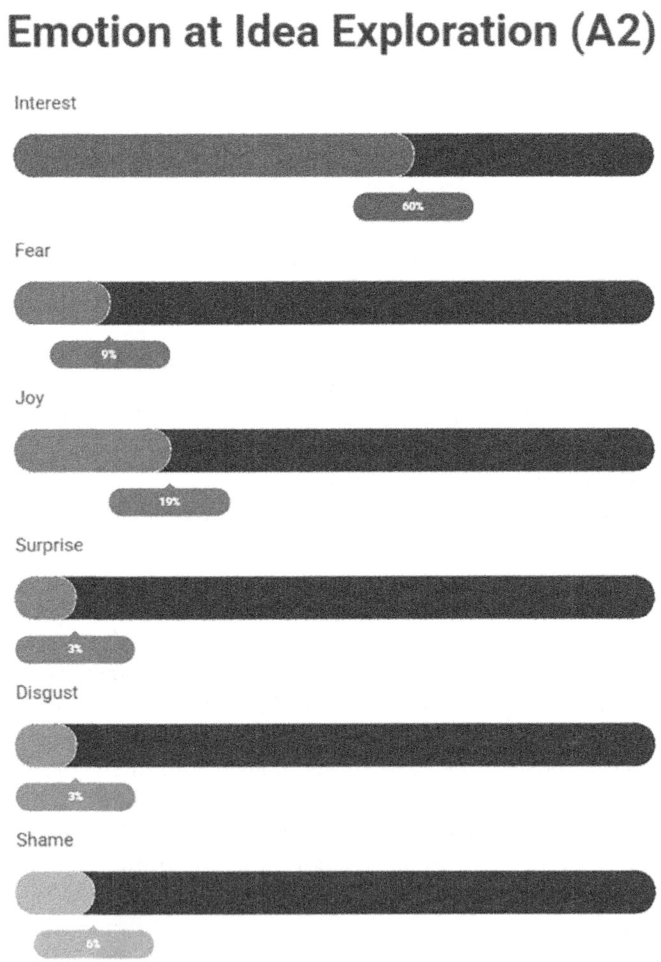

FIGURE 5.7 Participants' interest in idea exploration during the design process

with anger; specifically, fury, outrage, irritation, hostility, resentment or violence. Ninty-nine per cent of the participants were able to express an emotion that was either fearful or anxious in nature. Specifically, these emotions consisted of anxiety, apprehension, nervousness, dread, fright and panic. Nineteen per cent of the participants expressed an emotion that is associated with joy; particularly, those who were joyful exhibited a range of emotions, including enjoyment, happiness, relief, bliss, delight, pride, excitement and ecstasy. Three per cent of the participants expressed a sense of surprise or astonishment. In particular, this group of participants expressed emotions that included shock, amazement and wonder. Another 3% of the participants expressed a sense of disgust. More specifically, the negative emotions expressed by this group of participants included contempt, disdain, scorn, aversion, distaste and revulsion. Finally, 6% of the participants expressed a sense of shame; specifically, these emotions included guilt, embarrassment, remorse, regret and contrition.

B. Exploration and personal experience

During the design process, exploration and personal experience influence its aims and significance.

B1) I explored parts of the subject on my own (Figure 5.8).

The participants' exploration and personal experience influenced the purpose and significance of the design process. In this regard, 62.5% of the participants strongly agreed that they explored aspects of the subject by themselves, rating the statement 7 or higher, while 37.5% rated their exploration level between 5 and 4 and 0% rated their exploration level as 3 or lower.

B2) Which kind of emotion was evoked?

- aroused, astonished, stimulated, surprised, active
- enthusiastic, elated, excited, euphoric, lively, peppy

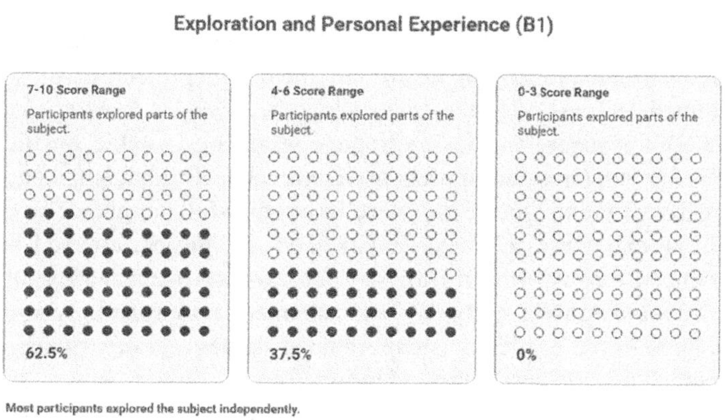

FIGURE 5.8 Participants' ratings of their exploration and personal experience

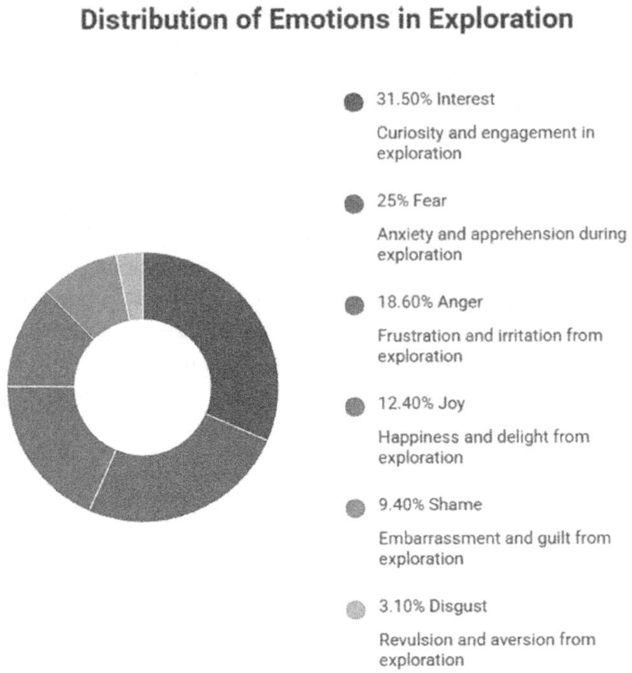

FIGURE 5.9 Participants' emotions evoked by the exploration and personal experience during the design process

- happy, delighted, glad, cheerful, warm-hearted, pleased
- relaxed, content, at rest, calm, serene, at ease
- quiet, tranquil, still, idle, passive
- dull, tired, drowsy, sluggish, bored, droopy
- unhappy, miserable, sad, grouchy, gloomy, blue
- distressed, annoyed, fearful, nervous, jittery, anxious (Figure 5.9)

The participants were asked to report the emotions that arose from their experience of exploring and being personally involved with the design project. Of those emotions, 31.5% expressed feelings of 'interest' (acceptance, friendly, trusting, kind, loving, devoted); 18.6% felt anger ('furious', outraged, 'wrathful', irritable, hostile, resentful, violent); 25% felt fear (anxious, apprehensive, nervous, afraid, frightened, panicked); 12.4% expressed joy (enjoyment, happy, relieved, blissful, delighted, proud, excited, ecstatic) and no one expressed surprise (shocked, astonished, amazed, wonderful); 3.1% experienced disgust (disdainful, scorned, averse, disgusted, repulsed); 9.4% experienced shame (guilty, embarrassed, ashamed, remorseful, sorry, regretful); and finally, none of the participants experienced sadness (grief-stricken, sorrowful, gloomy, melancholy, hopeless, lonely, depressed).

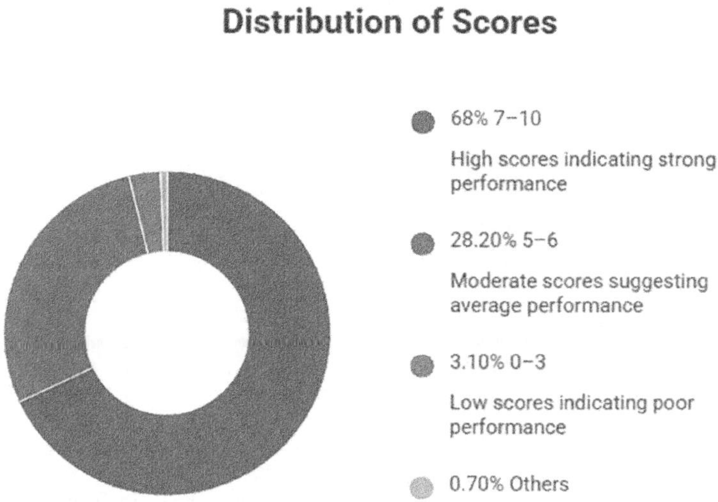

FIGURE 5.10 Participants' ratings of testing out their own ideas during the idea exploration phase

B3) I was able to learn by testing my own ideas (Figure 5.10).

Testing out my own ideas affected how much they enjoyed and experienced fulfilment from the design process. Here, 68% of the participants strongly agreed that they explored aspects of the subject by themselves, indicating a score of 7 or higher, while 28.2% rated their level of exploration between 5 and 6; and 3.1% rated their exploration level as 3 or lower.

B4) Which kind of emotion was evoked?

- aroused, astonished, stimulated, surprised, active
- enthusiastic, elated, excited, euphoric, lively, peppy
- happy, delighted, glad, cheerful, warm-hearted, pleased
- relaxed, content, at rest, calm, serene, at ease
- quiet, tranquil, still, idle, passive
- dull, tired, drowsy, sluggish, bored, droopy
- unhappy, miserable, sad, grouchy, gloomy, blue
- distressed annoyed, fearful, nervous, jittery, anxious (Figure 5.11)

Participants who found this course to be both challenging and exciting indicated that they experienced a variety of emotional responses. Only 18.8% expressed an interest in the course content. This interest was described as acceptance, friendly, trusting, kind, loving and devoted. Participants did not report feeling angry with respect to the course. The specific aspects of anger that were excluded are fury, outrage, irritation, hostility, resentment and/or violence, and 21.9% of participants

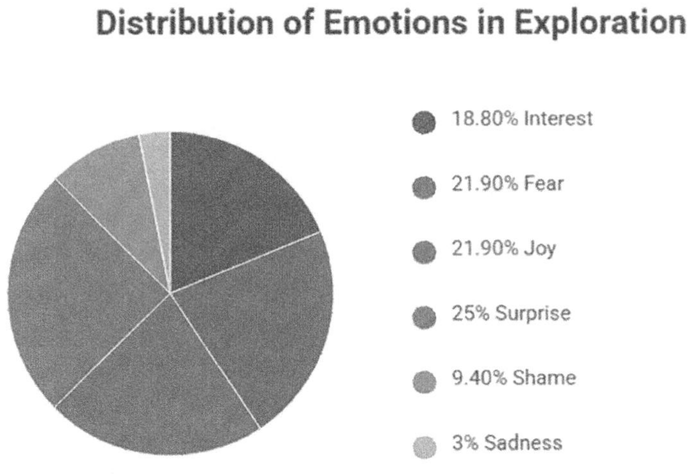

FIGURE 5.11 Participants' emotions evoked while learning by trying your own ideas

experienced fear during the completion of the course. Fear is defined as anxiety, apprehension, nervousness, dread, fright and/or panic. As with fear, 21.9% of participants also reported feelings of joy while completing the course. Joy is defined as enjoyment, happy, relieved, blissful, delighted, proud, excited and ecstatic. Twenty-five per cent of participants reported experiencing surprise at some time during the course. Surprise is defined as shock, astonishment, amazement and/or wonder. No participant identified disgust with respect to their feelings about the course. Disgust is defined as contempt, disdain, scorn, aversion, distaste and/or revulsion, and 9.4% of participants reported feelings of shame related to the course. Shame is defined as guilt, embarrassed, remorseful, regretful and/or contrite. Three per cent of participants reported experiencing feelings of sadness while taking the course. Sadness is defined as grief, sorrowful, gloomy, melancholy, despondent, lonely and depressed.

C. Challenge

The challenge level the participants experienced during the course affected how much enjoyment and fulfilment they experienced during the design process.

C1) The course was challenging in a stimulating way (Figure 5.12).

The challenge level of the idea exploration phase affected the degree of enjoyment and fulfilment the participants experienced during the design process. In this regard, 68.8% of the participants strongly agreed that they explored parts of the subject by themselves, indicated by a score of 7 or higher, while 46.9% rated their interest level between 5 and 6 and 9.4% rated their interest level as 3 or lower.

C2) Which kind of emotion was evoked?

- aroused, astonished, stimulated, surprised, active
- enthusiastic, elated, excited, euphoric, lively, peppy

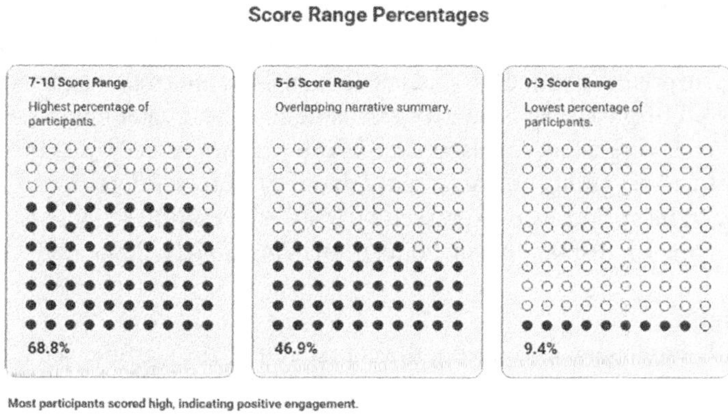

FIGURE 5.12 Participants' ratings of the challenge level of the course

- happy, delighted, glad, cheerful, warm-hearted, pleased
- relaxed, content, at rest, calm, serene, at ease
- quiet, tranquil, still, idle, passive
- dull, tired, drowsy, sluggish, bored, droopy
- unhappy, miserable, sad, grouchy, gloomy, blue
- distressed, annoyed, fearful, nervous, jittery, anxious (Figure 5.13)

When people perceived the course as both difficult and engaging, they experienced a variety of emotions, of which 21.9% expressed an emotion of interest that included acceptance, friendly, trustworthy, kind, loving, devoted, etc.; 25% expressed fear (fearful) – fear is defined as: anxiety, apprehensive, nervous, fearful, frightened,

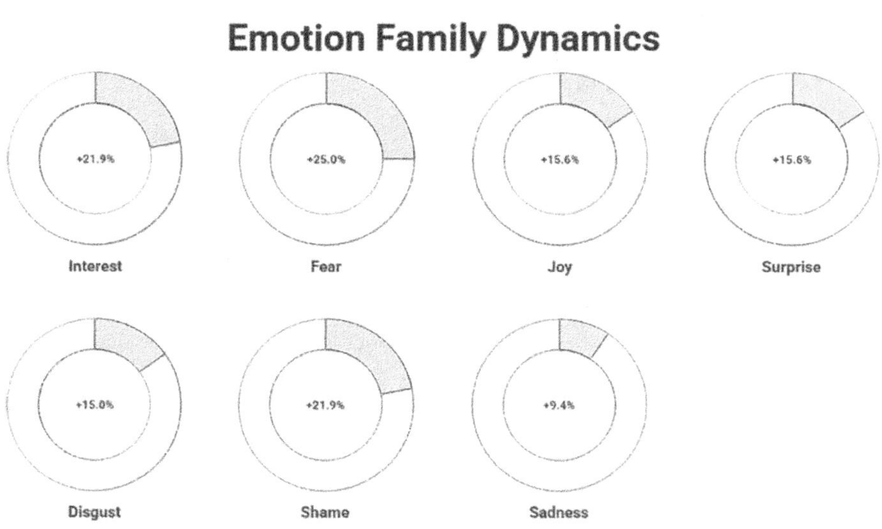

FIGURE 5.13 Emotions participants observed being evoked by challenges during the idea exploration phase

panicked, etc.; 15.6% expressed joy (joyous), joy is defined as: enjoyable, happy, relieved, blissful, delighted, proud, excited, ecstatic, etc.; 15.6% expressed surprise (surprised); surprise is defined as: shocked, astonished, amazed, wonder-struck, etc.; 15% expressed disgust (disgusted), disgust is defined as: contemptuous, disdainful, scornful, averse, disgusted, repulsed, etc. Shame was expressed by 21.9% (shamed), shame is defined as: guilty, embarrassed, ashamed, remorseful, regretful, repentant, etc. Finally, sadness (sad) was expressed by 9.4% (sorrowful), sadness is defined as: grieving, sorrowful, dismal, melancholic, despairing, lonely, depressed, etc.

D. Belonging

Feelings of belonging evoked during the course reflect the significance and satisfaction of the design process.

D1) I experienced a sense of togetherness with others who were completing the course (Figure 5.14).

The degree of belonging they experienced during the course (reflecting the significance of and level of satisfaction with the design process) allowed participants to understand what they were meant to achieve. Among the participants, 75% strongly agreed that they explored aspects of the subject by themselves, as indicated by a score of 7 or higher; 25% rated their exploration level between 4 and 6 and 0% rated their exploration level as 3 or lower.

D2) Which kind of emotion was evoked?

- aroused, astonished, stimulated, surprised, active
- enthusiastic, elated, excited, euphoric, lively, peppy
- happy, delighted, glad, cheerful, warm-hearted, pleased
- relaxed, content, at rest, calm, serene, at ease
- quiet, tranquil, still, idle, passive

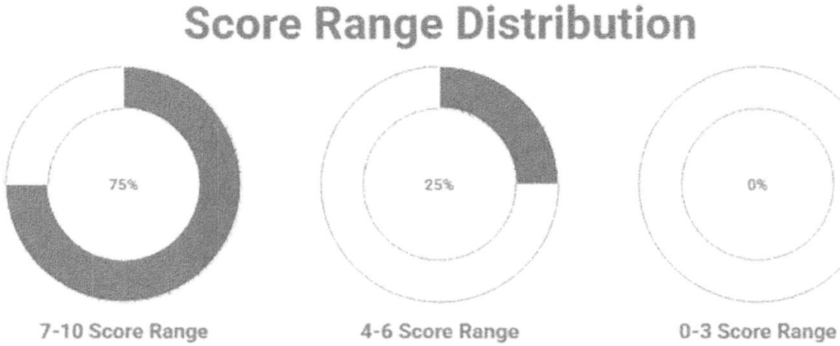

FIGURE 5.14 Participants' ratings concerning the degree of belonging they experienced during the course, reflecting the significance level of satisfaction with the design process

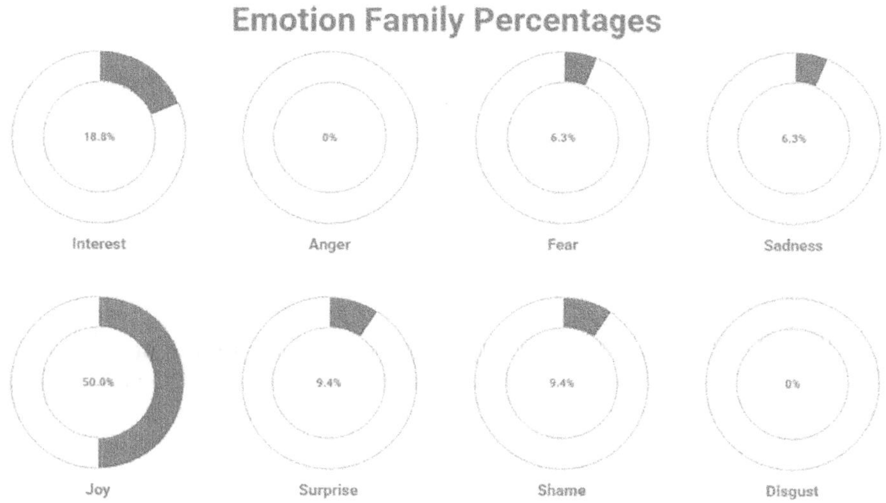

FIGURE 5.15 Participants' emotions, evoked by the togetherness with others in the course

- dull, tired, drowsy, sluggish, bored, droopy
- unhappy, miserable, sad, grouchy, gloomy, blue
- distressed, annoyed, fearful, nervous, jittery, anxious (Figure 5.15)

Participants responded with many different feelings when asked how they felt when thinking of a sense of togetherness with others that was developed during the course. Interest was felt by 18.8%. This interest included being accepted, friendly, trusting, kind, loving, devoted and other positive feelings towards people who are close to them. Not one participant felt angry. Anger is defined as: fury, outrage, wrath, irritability, hostility, resentment or violent feelings towards another person. Fear was the feeling of 6.3% of the participants. These fears were: anxiety, apprehension, nervousness, dread, fright and panic. The same percentage of participants (6.3%) felt sad. Sadness includes grief, sorrow, gloom, melancholy, despair, lonely, and depression. A total of 50% of the participants said they experienced joy. Joy encompasses enjoyment, happy, relieved, blissful, delighted, proud, excited, ecstatic, and other similar positive feelings. A total of 9.4% of the participants indicated that they felt surprised. Surprise can include shock, astonished, amazed and wonderful feelings towards something. Shame was felt by 9.4% of the participants. Shame includes guilty, embarrassed, chagrined, remorseful, regretful and contrite feelings. Disgust was experienced by none of the participants. Disgust includes contemptuous, disdainful, scorned, aversive, disgusted and repulsed feelings towards people.

D3) The atmosphere in the course was open and inclusive (Figure 5.16).

The opportunity to learn by trial enabled participants to understand what they were expected to achieve. Here, 78.2% of the participants strongly agreed that they

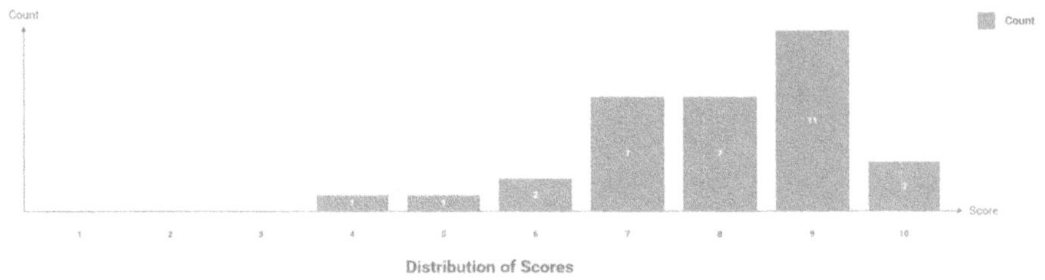

FIGURE 5.16 Participants' ratings of the atmosphere in the course was open and inclusive.

explored aspects of the subject on their own, reflected by a score of 7 or higher; 12.5% rated their exploration level between 4 and 6; and 0% rated their exploration level as 3 or lower.

D4) Which kind of emotion was evoked?

- aroused, astonished, stimulated, surprised, active
- enthusiastic, elated, excited, euphoric, lively, peppy
- happy, delighted, glad, cheerful, warm-hearted, pleased
- relaxed, content, at rest, calm, serene, at ease
- quiet, tranquil, still, idle, passive
- dull, tired, drowsy, sluggish, bored, droopy
- unhappy, miserable, sad, grouchy, gloomy, blue
- distressed, annoyed, fearful, nervous, jittery, anxious (Figure 5.17)

The participants were asked to identify the emotions that were evoked by the open and inclusive atmosphere of the course. Here, 34.4% of the participants indicated

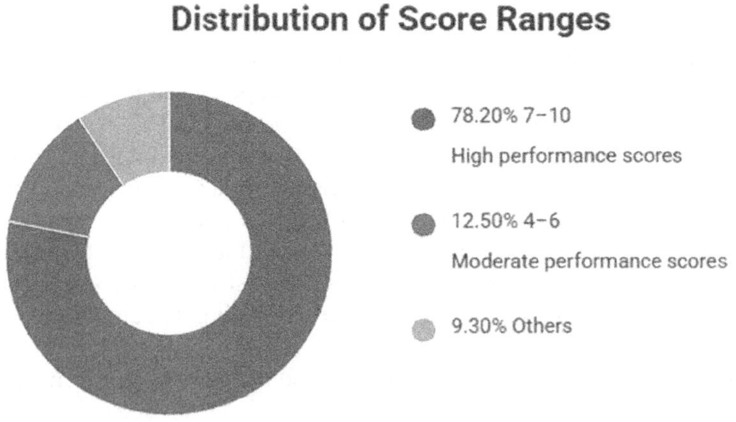

FIGURE 5.17 Participants' emotions evoked by the open and inclusive atmosphere of the course

emotions related to interest (including acceptance, friendliness, trust, kindness, affection, love and devotion), 0% denoted feelings of anger (including fury, outrage, wrath, irritability, hostility, resentment and violence), 0% experienced feelings related to fear (including anxiety, apprehension, nervousness, dread, fright and panic), 46.9% denoted feelings of joy (including enjoyment, happiness, relief, bliss, delight, pride, excitement and ecstasy), 12.5% experienced feelings of surprise (including shock, astonishment, amazement, surprise and wonder), 0% experienced disgust (including contempt, disdain, scorn, aversion, distaste and revulsion), 3.1% experienced shame (including guilt, embarrassment, chagrin, remorse, regret and contrition) and 3.1% experienced feelings of sadness (including grief, sorrow, gloom, melancholy, despair, loneliness and depression).

5.2.3.3 Part III: Comprehensibility – The Cognitive Level

The levels of productivity and complexity of the design process will reflect the design process' comprehensibility. Regarding the design process in the final poster project, its comprehensibility was reflected by the following aspects: clear objectives, an organised approach and an understanding of the subject.

A. Clear objectives and organisation

During the design process, clarity of the objectives and organisation were closely linked with design purpose and design productivity.

A1) The intended learning outcomes helped me understand what I was expected to achieve (Figure 5.18).

The ability to learn by trial helped participants understand what they were expected to achieve. In this instance, 75% of the participants strongly agreed that they explored parts of the subject by themselves, as reflected by a score of 7 or higher; 25% rated their interest between 4 and 6.0 and 0 rated their interest level as 3 or lower.

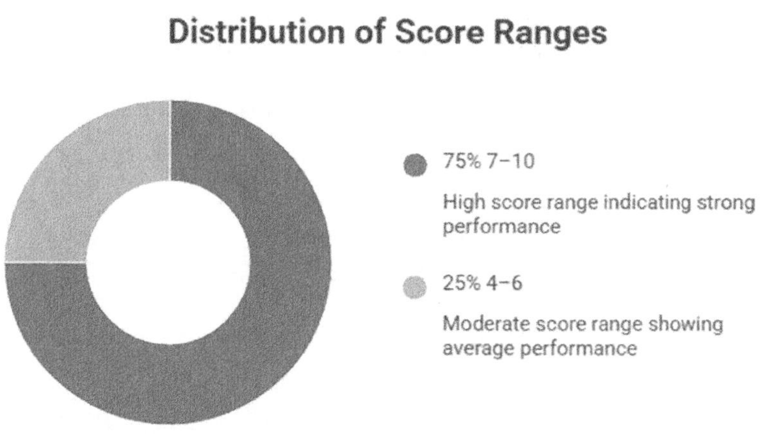

FIGURE 5.18 Participants' ratings of their ability to achieve the intended learning outcomes

A2) Which kind of emotion was evoked?

- aroused, astonished, stimulated, surprised, active
- enthusiastic, elated, excited, euphoric, lively, peppy
- happy, delighted, glad, cheerful, warm-hearted, pleased
- relaxed, content, at rest, calm, serene, at ease
- quiet, tranquil, still, idle, passive
- dull, tired, drowsy, sluggish, bored, droopy
- unhappy, miserable, sad, grouchy, gloomy, blue
- distressed annoyed, fearful, nervous, jittery, anxious (Figure 5.19)

The emotions that respondents felt when considering their potential learning outcomes varied. A total of 40.6% of respondents expressed an emotional interest in this outcome and included acceptance, friendliness, trust, kindness, love, affection and devotion among these feelings. There was no respondent who indicated a feeling of anger as this emotion is defined to include curiosity, outraged-ness, irascibility, wrath, enmity, rancour, ire, etc. Approximately 18.8% of respondents indicated that they would experience fear when considering their potential learning outcome and included the feelings of anxiety, fearfulness, apprehensiveness, dread, fright, panic, etc., within this category. This percentage also corresponds with those who will experience joy or pleasure when considering their potential learning outcome and included within this group are enjoyment, happiness, relief, bliss, satisfaction, pride, enthusiasm and ecstasy, and 15.6% of respondents indicated that they will be surprised when considering their potential learning outcome and this group includes the feelings of shock, amazement, awe, astonishment, etc. There were no feelings of disgust reported by respondents and the feelings of disgust are defined to include feelings of loathing, contempt, disdain, scorn, aversion, distaste and revulsion. And 3.1% of respondents reported experiencing the feelings of shame and these feelings of shame are defined to include feelings of guilt, embarrassment, chagrin, remorse, regret and contrition. Approximately 3.1% of respondents reported experiencing the feelings of sadness and these feelings of sadness are defined to include feelings of grief, sorrow, gloom, melancholy, despair, loneliness and depression.

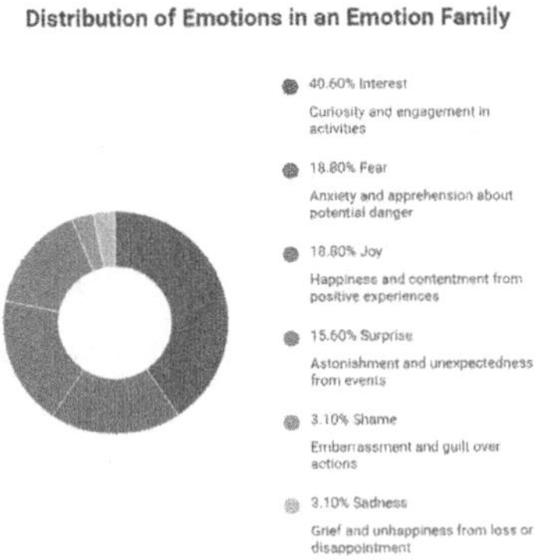

FIGURE 5.19 Participants' emotions evoked by the expected learning outcomes

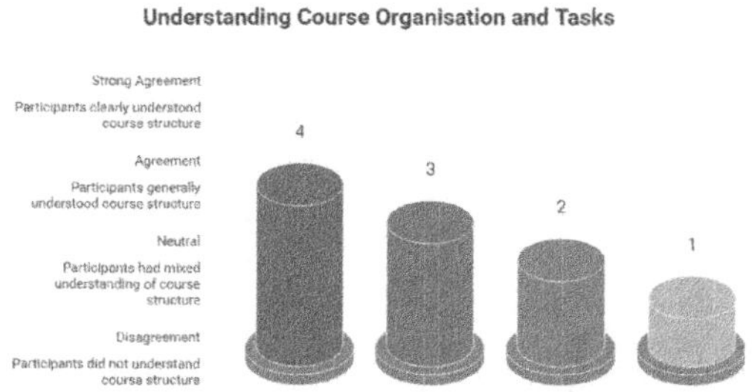

FIGURE 5.20 Participants' ratings of their ability to learn the concepts related to emotional concerns and design in the course

A3) I understood how the course was organised and what I was expected to do (Figure 5.20).

A4) Which kind of emotion was evoked?

- aroused, astonished, stimulated, surprised, active
- enthusiastic, elated, excited, euphoric, lively, peppy
- happy, delighted, glad, cheerful, warm-hearted, pleased
- relaxed, content, at rest, calm, serene, at ease
- quiet, tranquil, still, idle, passive
- dull, tired, drowsy, sluggish, bored, droopy
- unhappy, miserable, sad, grouchy, gloomy, blue
- distressed annoyed, fearful, nervous, jittery, anxious (Figure 5.21)

The participants were asked to present the emotions that were evoked by their understanding of how the course was organised and what they were expected to do. Here, 53.1% of the participants denoted feelings of interest (including acceptance, friendliness, trust, kindness, affection, love and devotion), 0% denoted feelings of anger (including fury, outrage, wrath, irritability, hostility, resentment and violence), 6.3% denoted feelings of fear (including anxiety, apprehension, nervousness, dread, fright and panic), 15.6% denoted feelings of joy (including enjoyment, happiness, relief, bliss, delight, pride, excitement and ecstasy), 12.5% denoted feelings of surprise (including shock, astonishment, amazement, surprise and wonder), 0% denoted feelings of disgust (including contempt, disdain, scorn, aversion, distaste and revulsion), 12.5% denoted feelings of shame (including guilt, embarrassment, chagrin, remorse, regret and contrition) and 0% denoted feelings of sadness (including grief, sorrow, gloom, melancholy, despair, loneliness and depression).

B. Understanding the subject

B1) I was able to learn from concrete and relatable examples (Figure 5.22).

The ability to learn from concrete examples influenced the participants' motivation to create new ideas. Accordingly, 71.9% of the participants strongly agreed that they

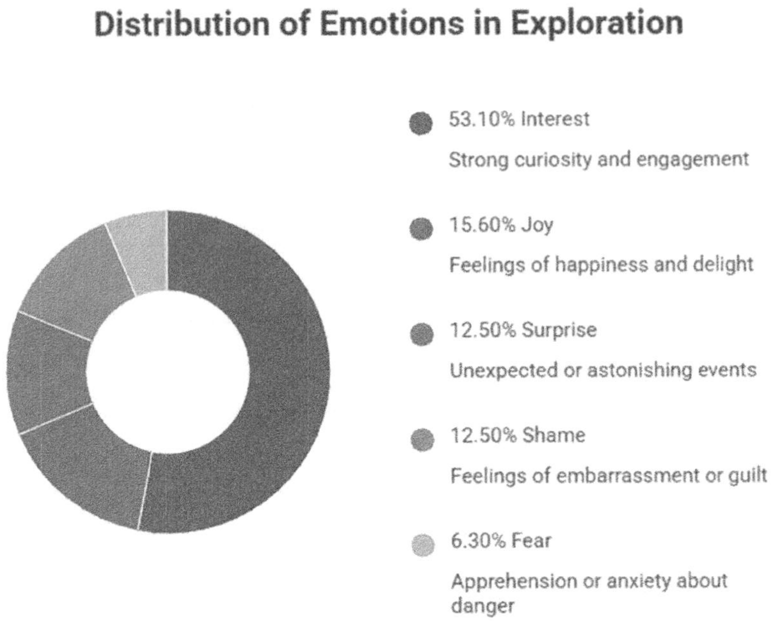

FIGURE 5.21 Participants' emotions evoked by their ability to understand how the course was organised

explored aspects of the subject by themselves, as reflected by a score of 7 or higher, while 25% of the participants rated their interest between 5 and 6 and 3.1% rated their interest level as 3 or lower.

B3) Which kind of emotion was evoked?

- aroused, astonished, stimulated, surprised, active,
- enthusiastic, elated, excited, euphoric, lively, peppy
- happy, delighted, glad, cheerful, warm-hearted, pleased
- relaxed, content, at rest, calm, serene, at ease
- quiet, tranquil, still, idle, passive
- dull, tired, drowsy, sluggish, bored, droopy
- unhappy, miserable, sad, grouchy, gloomy, blue
- distressed annoyed, fearful, nervous, jittery, anxious (Figure 5.23)

The participants were asked to present the emotions that were evoked by their understanding of the way the course was organised and what they were expected to do. Here, 53.1% of the participants denoted feelings of interest (including acceptance, friendliness, trust, kindness, affection, love and devotion), 0% denoted feelings of anger (including fury, outrage, wrath, irritability, hostility, resentment and violence), 6.3% denoted feelings of fear (including anxiety, apprehension, nervousness,

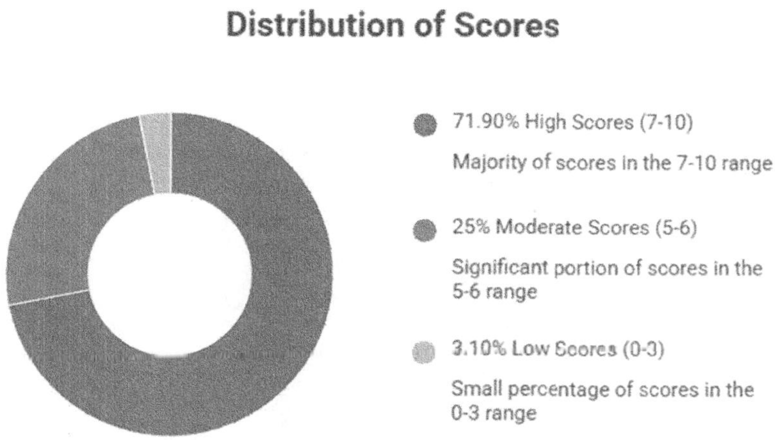

FIGURE 5.22 Participants' ratings related to learning from concrete examples

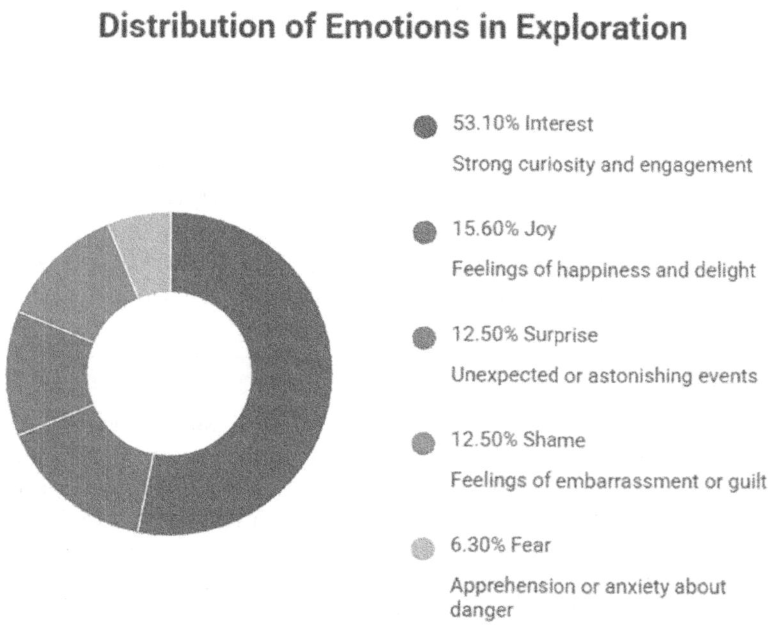

FIGURE 5.23 Participants' emotions evoked by their understanding of how the course was organised and what they were expected to do

dread, fright and panic), 15.6% denoted feelings of joy (including enjoyment, happiness, relief, bliss, delight, pride, excitement and ecstasy), 12.5% denoted feelings of surprise (including shock, astonishment, amazement, surprise and wonder), 0% denoted feelings of disgust (including contempt, disdain, scorn, aversion, distaste and revulsion), 12.5% denoted feelings of shame (including guilt, embarrassment,

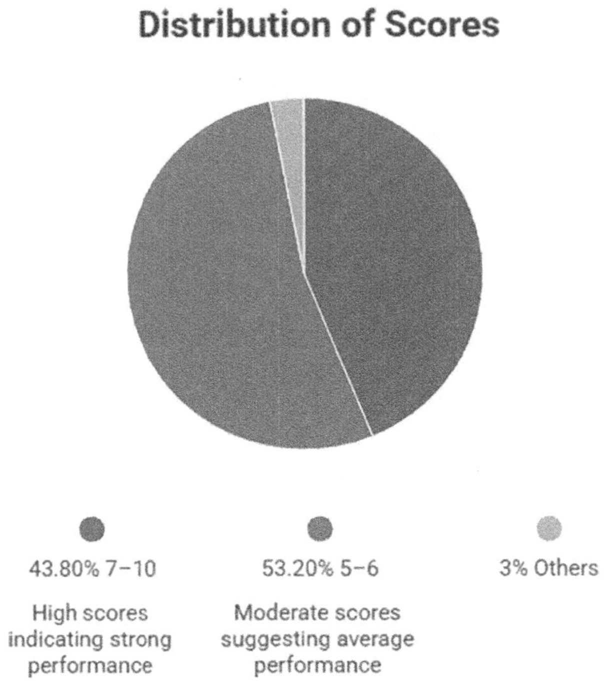

FIGURE 5.24 Participants' ratings of their understanding of key concepts

chagrin, remorse, regret and contrition) and 0% denoted feelings of sadness (including grief, sorrow, gloom, melancholy, despair, loneliness and depression).

B4) Understanding key concepts is paramount (Figure 5.24).

The ability to learn and apply new methods influenced participants' motivation to create new ideas. Here, 43.8% of the participants strongly agreed that they explored aspects of the subject by themselves, reflected by a score of 7 or higher, while 53.2% rated their interest between 5 and 6 and 0% rated their interest level as 3 or lower.

B5) What emotions were evoked?

- aroused, astonished, stimulated, surprised, active
- enthusiastic, elated, excited, euphoric, lively, peppy
- happy, delighted, glad, cheerful, warm-hearted, pleased
- relaxed, content, at rest, calm, serene, at ease
- quiet, tranquil, still, idle, passive
- dull, tired, drowsy, sluggish, bored, droopy
- unhappy, miserable, sad, grouchy, gloomy, blue
- distressed annoyed, fearful, nervous, jittery, anxious (Figure 5.25)

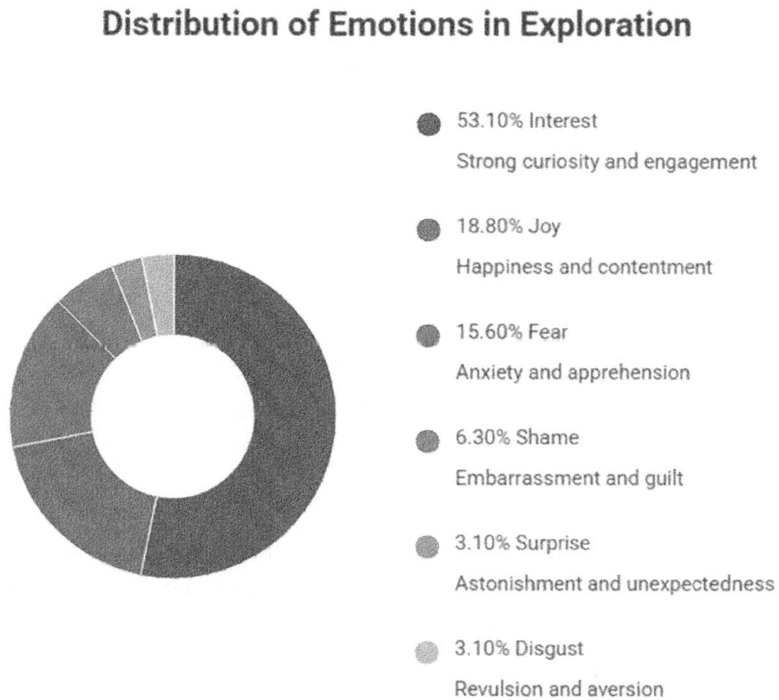

FIGURE 5.25 Participants' emotions evoked by understanding key concepts

The participants were asked to present the emotions that were evoked by understanding the key concepts of the project. Here, 53.1% of the participants stated that feelings of interest were evoked (including acceptance, friendliness, trust, kindness, affection, love and devotion), 0% denoted feelings of anger (including fury, outrage, wrath, irritability, hostility, resentment and violence); 15.6% denoted feelings of fear (including anxiety, apprehension, nervousness, dread, fright and panic); 18.8% denoted feelings of joy (including enjoyment, happiness, relief, bliss, delight, pride, excitement and ecstasy); 3.1% denoted feelings of surprise (including shock, astonishment, amazement, surprise and wonder); 3.1% denoted feelings of disgust (including contempt, disdain, scorn, aversion, distaste and revulsion); 6.3% denoted feelings of shame (including guilt, embarrassment, chagrin, remorse, regret and contrition); and 0% denoted feelings of sadness (including grief, sorrow, gloom, melancholy, despair, loneliness and depression).

5.2.3.4 Part IV: More Questions About Special Moments in the Design Process

The following questions guided the participants to think about special moments, the overall significance and the design productivity in the design process of the final poster design project.

1. What were some of the most interesting discoveries I made while working on this project? About the problem? About myself? About others?
    - Different layout designs are interesting.
    - The style of Armin Hofmann and his work encouraged me a lot.
    - My ability to be imaginative.
    - Observing the artwork of various artists.
    - I found that typeface style can really affect the poster design.
    - The use of AI.
    - Developing many ideas for the poster.
    - I had some interesting ideas while creating my poster, and the process was fun and enjoyable.
    - Reflection on how we use the same technology.
    - My own abilities.
    - Designing the poster.
    - Trying to use AI to implement the idea.
    - Understanding myself better.
    - What I thought is not the same as what I did.
    - I discovered many examples and began observing the posters around me more. I realised the importance of composition in designing the poster and found it interesting that I could understand some ('Oh, so that's why it's like that').

      During this project, however, I learned that I do not have very good time management skills and will work on developing these. In addition, I found that digital painting and digital design (manipulating fonts) using software requires additional learning.
    - Know more about yourself.
    - How to mix technology and nature.
    - Everyone has different great ideas.
    - I learned that the more I try, the more I will be able to explore.
    - I observed many good designs created by my classmates, and I discovered many interesting functions of AI.
    - I try to get different cultures' backgrounds and stories. I wouldn't have read the information before.
    - I can design a poster by myself.
    - The discussions with others.
    - How I can use the basic visual element to express my thoughts and ideas to others.
    - I could learn more software skills (Adobe) by myself while completing the assignment, and the process of doing the assignment was interesting.
    - Finding myself to be creative.
    - I found that I appreciate a simple design.
    - Technique-related problems.
    - I can design a poster by myself.
    - Finding myself to be creative.
    - My own capability.
    - Finding myself to be creative.

2a. What was my most challenging moment in the entire design process and why?
- Visualising the ideas in my brain. I have the idea but not the ability.
- The use of Photoshop, Adobe Illustrator, etc.
- Creating from nothing.
- Learning how to use different software on our own.
- Making the design clear and interesting.
- Idea discovery.
- The digital poster!
- It was hard to work with the design and drawing software for a beginner.
- Do the digital final outcome
- Visualisation of ideas, as I am not experienced.
- The mid-term project.
- Developing different kinds of ideas.
- Achieving what I imagined. My skills are not sufficient to achieve the standard that I expected.
- Being without design ideas.
- The use of software, because I was not experienced in this area.
- Not enough design ideas.
- How to apply the visual element to the project.
- Developing 30 thumbnails, because it's difficult to think up that many ideas.
- Designing the poster digitally.
- Using AI and drawing the thumbnails. It was difficult to develop my idea in the beginning, and I did not have many ideas for drawing the thumbnails.
- Drawing out the poster, because I am not really familiar with digital drawing.
- Thinking about designing a poster from beginning to end.
- Developing the theme for the poster, based on the artist we studied.
- Turning the graphic into a digital image, since I have no experience in using editing software.
- Drawing anything during the course was the hardest part.
- I could not present my idea completely due to some technical problems.
- Colouring, because I think without colour is more beautiful.
- My skill in Illustrator and the first meeting.
- I could not present my idea completely due to some technical problems.
- Visualisation of ideas, as I am not experienced.
- I could not present my idea completely due to some technical problems.

2b. What emotions were evoked during the most challenging moments of the design process?
- aroused, astonished, stimulated, surprised, active
- enthusiastic, elated, excited, euphoric, lively, peppy
- happy, delighted, glad, cheerful, warm-hearted, pleased
- relaxed, content, at rest, calm, serene, at ease

FIGURE 5.26 Participants' feedback on the question 'Which kind of emotion was evoked during the most challenging moment in the design process?'

- quiet, tranquil, still, idle, passive
- dull, tired, drowsy, sluggish, bored, droopy
- unhappy, miserable, sad, grouchy, gloomy, blue
- distressed annoyed, fearful, nervous, jittery, anxious (Figure 5.26)

3a. How do I feel my solution (the poster design) relates to real-world situations and problems?
- Relates metaphorically.
- Not sure.
- Don't know.
- It is okay for a junior like me.
- Very happy and joyful.
- Not very related.
- Relaxed.
- Satisfied.
- Quite good.
- Practical.
- Not the best, but I tried.
- I feel the poster is too simple to relate to a real-world problem, but I think it can show the situation.
- Ask help from others and find more references.

- I don't know.
- Having to cope with the disappointment or redoing work again and again can occur in many real-life scenarios, as can … working to meet a deadline. How I dealt with these situations will surely come in handy in the real world.
- I feel free.
- Not bad.
- My poster can remind me to do things my own way.
- The solution that I came up with was to push ourselves even when we were feeling down. I think this solution can be applied to real-life situations and problems.
- I think the problem that I mentioned is common in Hong Kong. The pace of living in Hong Kong is fast; we are always affected by other things, such as when we play, we will receive messages from our boss or company. So, we can't really enjoy ourselves and feel relaxed when we have a rest. Therefore, I hope that my poster can clearly show enjoyment and relaxation.
- I used a full day to draw the poster, and I felt a little bit bothered midway, but I tried to calm down and continue the work.
- Not bad.
- I think the situations and problems affected my design a bit negatively.
- It is really helpful. And it is beneficial for our daily life.
- How to execute my idea.
- Unsatisfied.
- The project was similar to a real-world situation, and I will try to design more and choose the best work.
- It related through colours and shapes.
- Not bad.
- Unsatisfied.
- Practical.
- Unsatisfied.

3b) Which kind of emotion was evoked during the most challenging moments of the design process?

- aroused, astonished, stimulated, surprised, active
- enthusiastic, elated, excited, euphoric, lively, peppy
- happy, delighted, glad, cheerful, warm-hearted, pleased
- relaxed, content, at rest, calm, serene, at ease
- quiet, tranquil, still, idle, passive
- dull, tired, drowsy, sluggish, bored, droopy
- unhappy, miserable, sad, grouchy, gloomy, blue
- distressed annoyed, fearful, nervous, jittery, anxious (Figure 5.27)

FIGURE 5.27 Participants' feedback on the question 'Which kind of emotion was evoked during the most challenging moment in the design process?'

5.3 RESEARCH STUDY 2

5.3.1 Phase 1: Examination of the Effectiveness of Individual Proposed Principles

In the fourth stage, another design project was assigned. Participants were invited to record their emotions using an emotion-tracking application while conducting research and engaging in discussions and idea generation. The objective of this stage was to record emotional changes visually and examine the effectiveness of individual proposed principles.

5.3.1.1 Input Work Effort

Concerning the design process of the final poster for IDDA A200F, Foundation of Visual Communication BFA (Hons), Imaging Design and Digital Art, it was essential to understand the degree of effort participants applied during a week. After collecting the participants' feedback for Research Study 1 and Research Study 2, the data were gathered and presented using the following charts. Participants' work efforts in Research Study 1 and Research Study 2 were similar. In other words, the workload was similar in both studies (Figure 5.28).

The emotional changes participants experienced during the design process were collected using an emotion-tracking application. Based on the feedback from participants, five emotions were identified. Pleasure (35%) received the most votes, followed by interest (25%), elevated mood (19%), anxiety (10.5%), and irritability (10.5%). The emotion-tracking application successfully tracked participants' emotional changes, which helped them to think about how they could manipulate the design process in a future experience and how it would compare with the current one.

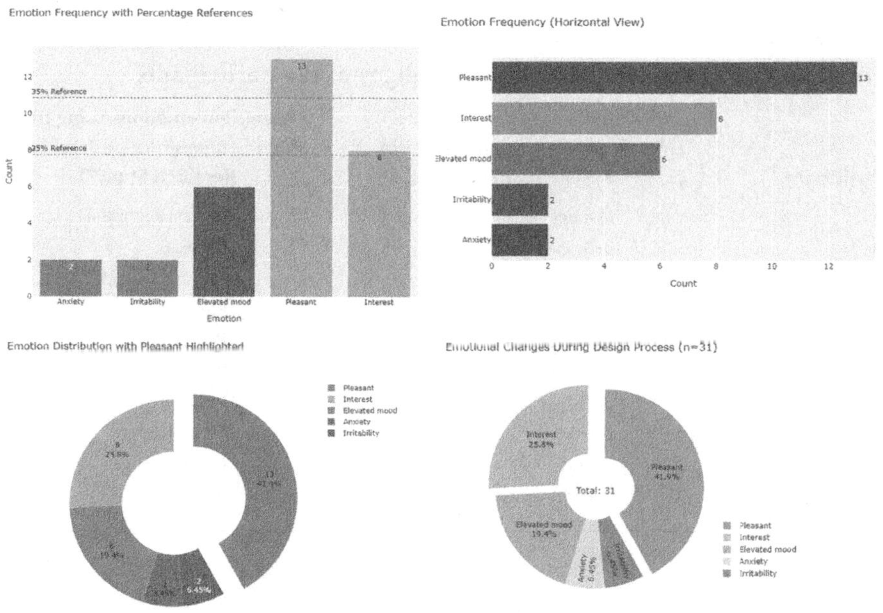

FIGURE 5.28 The emotions identified by the eMoods Bipolar Mood Tracker in Research Study 2

The participants identified their emotions using the eMoods Bipolar Mood Tracker while working on the assigned design task in Research Study 2. Here, 2 participants identified anxiety; 2 identified irritability; no participants identified a depressed mood; 6 identified an elevated mood; 13 identified pleasure; 8 identified interest; none of the participants identified feeling sad; and no uncategorised feelings were denoted (Table 5.3).

5.3.2 Phase 2: Focus Groups

The CAPTIV L7000-Premier emotion-tracking mobile application was used to conduct 45-minute interviews with participants in 14 groups to study their design process and experience as influenced by the effect of emotions. Participants were invited to talk about how they had used the suggested emotion and design principles in their own design work. The smartphone application's record of emotions was examined. The CAPTIV-Neurolab method, a system for recognising emotional reactions, was used in the focus group discussion to better understand participants' emotional shifts. According to Kimmel et al. (2024), the findings of the CAPTIV L7000-Premier showed an improvement in emotional responses, eye movement and self-control tasks. Several additional systems for tracking emotional reactions were compared to the CAPTIV L7000-Premier by Kimmel et al. (2024). Researchers have utilised the CAPTIV L7000-Premier to study the requirements of both adults and children.

In previous studies, research participants' mental systems' (i.e., mental analysis algorithms) benefited greatly from the use of the emotional reactions monitoring

TABLE 5.3
The (randomly extracted) emotional changes of ten participants

Participant	Identified emotions using the eMoods Bipolar Mood Tracker in Research Study 1	Identified emotions using the eMoods Bipolar Mood Tracker in Research Study 2
A	Anxiety	Elevated mood
B	Irritability	Interest
C	Depressed mood	Elevated mood
D	Elevated mood	Pleasure
E	Anxiety	Pleasure
F	Elevated mood	Interest
G	Anxiety	Interest
H	Irritability	Pleasure
I	Depressed mood	Interest
J	Depressed mood	Interest

approach. Emotional literacy is the ability to accurately convey oneself via facial expressions and body language (Noffke & Somekh, 2009). To test the suggested concepts individually, the efficacy of the CAPTIV L7000-Premier system provided scientific data for each design step.

5.3.2.1 Reflection on the Design Process

While capturing frames, CAPTIV L7000-Premier was implemented to detect faces (as shown in Figure 5.29). The CAPTIV data source was used for facial expression

FIGURE 5.29 CAPTIV L7000-Premier was used to detect faces

FIGURE 5.30 An example of participants' work (designed by Participant 1)

analysis. This method overcame typical problems in facial expression recognition in a focus group, including the need for appropriate database settings and the difficulty of standardising results. To ensure the reliability and validity of CAPTIV L7000-Premier in identifying the participants' facial expressions, a literature study and comparisons of the experimental settings were conducted. The participants provided their input during interviews, discussed possible issues with the design process and proposed guiding principles (Figure 5.29).

FIGURE 5.31 An example of participants' work (designed by Participant 2)

5.3.3 Phase 3: Individual Interviews and the Reflective Reviews in Digital Lablog

A 20-minute interview and a lablog review were conducted with the top and bottom three achievers in academic results evaluation among the participants to better understand how emotions impacted their design process. The researcher recorded the interview and transcribed it to enable coding throughout this phase. Qualitative and quantitative analyses of the coding were conducted, and new coding systems were devised.

5.3.3.1 Reflective Review in Participants' Lablog

The lablog included a full record of the learning process and served as a space for trying new things, that is, a reflective journal and thinking tool (Figures 5.30 and 5.31).

6 Discussion

6.1 OVERVIEW

Interest in the connections between emotional concerns and design methods was prompted by the work of various design researchers in the 2000s. Their studies focused primarily on the perspectives of users. Along with logic and reason, the design process involves skills, knowledge and abilities of designers, as well as their responses and reactions to the external environment/stimuli they encounter. Few research studies have examined the emotional changes that junior design students experience during the design process, despite the fact that these studies offer crucial theoretical foundations. Junior design students, particularly, may encounter difficulties in comprehending information, particularly when dealing with designs that involve intricate human interactions and complexity. The interviews underscored the importance of incorporating emotional and human-centred considerations into the design processes of junior students, while also effectively managing these processes. Nevertheless, viable resolutions to these challenges have not yet been determined.

Bloom and Krathwohl (1956) classified learning goals as developing cognitive, emotional and psychomotor skills (Anderson & Krathwohl, 2001). Conceptual, procedural and metacognitive knowledge are characterised by the most basic to the most sophisticated knowledge functions: knowing, comprehending, applying, analysing, evaluating and producing. Bloom's taxonomy is commonly used by educators to set learning goals. Affective taxonomy is seldom used. Affective learning outcomes are more difficult to measure than cognitive results such as language recall (Krathwohl et al., 1964). Design, which emphasises emotive abilities, requires affective taxonomies.

Affective domains regulate emotions and internalise them. Internalisation refers to the consistent transformation of personal knowledge of the external environment and phenomena into a pervasive general attitude to life that impacts all the individual's reactions and activities (Krathwohl et al., 1964). The affective skill taxonomy presented by Krathwohl et al. categorises emotions (from basic to complicated) as receiving, reacting, valuing, organising and characterising. Each step of this model comprises sub-levels that outline the internalisation process.

Within the framework of human-centred design, Zoltowski et al. (2012) examined a taxonomy of emotional skill development. In their visual engineering study of undergraduate students who had engaged in a user-centred design process, the researchers wanted to understand how students at various levels perceive the human-centred design process. Their investigation found seven types of engagement within the realm of design process and user knowledge. Two groups reflect pre-engagement in any dimension. Two of the five remaining categories detail students' experiences without design or user demands. Thus, categories three through seven indicate student

DOI: 10.1201/9781003747345-6

progress within an empathic, human-centred design process. The pre-empathy and pre-design categories correspond to Krathwohl et al.'s (1964) assertion that pupils must be willing and able to interact with phenomena before the internalisation of emotional skills can occur. Although Zoltowski et al. (2012) did not cite Krathwohl et al.'s taxonomy, their five categories show agreement with the five levels suggested by the latter. Even more useful, it includes descriptions of design behaviour at each level, thereby not only showing that emotional skills develop throughout design education but also that this development is essential to design students' success.

Affective abilities improve with practice (Krathwohl et al., 1964; Zoltowski et al., 2012). Student behaviour and linguistic elicitation reveal growth phases. Both suggest that academic programmes may educate and measure emotional qualities such as empathy. In addition, it allows performance and behaviour-based skill evaluation. Existing research in design education indicates that this training is not occurring as effectively as required or is not being transferred to new problem sets, especially where multiple user groups are involved in various design projects. As such, pedagogical strategies capable of teaching emotional skills must be studied in the context of design education.

This chapter describes how the findings obtained in Research Studies 1 and 2 fulfil this study's objectives:

- To discover how junior design students could apply emotional components into their own design processes.
- To determine what emotional aspects allow junior design students to better govern their design processes.
- To establish guiding principles to assist junior design students in managing the emotional variations they experience throughout the design process.
- To assess the impact of these principles on helping junior design students to govern their emotional variability while designing.
- To research ways to encourage junior design students to utilise regulated emotional components as a means to affect and improve their design process.

6.2 RESEARCH STUDY 1

6.2.1 Phase 1: Investigation of the Emotional Elements in the Design Process

Research Study 1 contained enough information to answer Research Question 1: 'What types of emotions are recognised by junior design students while making decisions in their design process'. The subjects stated that they experienced emotions like: anxiety, irritability, depression, euphoria, pleasure and interest. However, since many of these students had never studied emotion awareness, many stated that their feelings fell into the category of 'uncategorised' because they did not know how to define their emotions while working through their design assignments. The subjects of Research Study 2 also indicated they felt the same type of emotions – e.g., anxiety, irritability, depression, euphoria, pleasure and interest. This was the first time the subjects had been instructed in how to categorise their emotions as part of a lecture before beginning work on the design assignment for this study; therefore, none classified their feelings as 'uncategorised', indicating an increase in emotional awareness.

Based on the lecture on observing emotions and the quantitative findings derived from Research Studies 1 and 2, an answer was provided for Research Question 3, 'How do junior design students' design experiences during the design process relate to emotion, and how is the influence of emotion connected to cognition in relation to emotion?' The emotions reported in Research Study 1 involved the designers' perceptions; however, they did not observe having been under the influence of the emotional changes by themselves.

The quantitative findings of Research Studies 1 and 2 provided an answer to Research Question 4, 'How can the proposed "Principles for Design and Emotion Concept Configuration in the Design Process" help them to manipulate their design process?' By comparing the emotions identified by the participants in Research Studies 1 and 2, it was observed that the ability to assist junior design students in controlling their emotions in the design process had been achieved. Compared to Research Study 1, students identified more positive emotions in Research Study 2. In Research Study 2, more participants identified pleasant and interested emotions and fewer participants identified anxiety, irritability, a depressed mood or an elevated mood. As they learned how to recognise their own emotions during the lecture before they worked on the design tasks assigned in Research Study 2, none of the participants identified 'uncategorised' emotions. Additionally, the junior design students' performance in managing their design processes was improved. Design outcomes that are generated through the design process employing the Principles for Design and Emotion Concept Configuration in the Design Process are thus expected to be optimised (Figure 6.1).

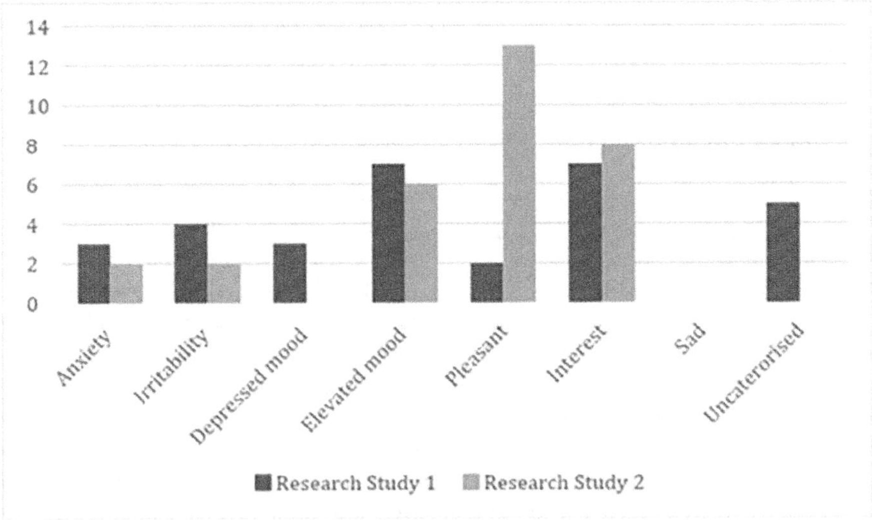

FIGURE 6.1 A comparison of the identified emotions in Research Studies 1 and 2

6.2.2 Phase 2: Introduction of the Proposed Principles of Controlling Emotions in Design

6.2.2.1 Approaches for Junior Design Students to Introduce Emotions into Their Design Processes

Approaches for junior design students to introduce emotions into their design processes were defined. Research on EI has identified that effective communication skills depend on the ability to negotiate and to negotiate with others. As previously stated, we live in a pluralistic society in which people hold different points of view. Therefore, in the process of negotiation, we may not fully agree with the views and behaviours of others, which may give rise to discontentment.

Some of the participants shared similar viewpoints. They stated that people may not be able to vocalise their dissatisfaction, which could damage the relationship between the two sides, so the process cannot progress as usual. They agreed that well-developed negotiation skills enable people to express not only their dissatisfaction from the heart but also constructive criticism related to their discontent and to negotiate for reasonable conditions for themselves while considering the interests of others, thus ensuring that both sides benefit. Diversity and teamwork without conflict are characteristics of modern successful design teams (Solmi, et al., 2020; Soto Hormazába, 2021).

According to a medical study, 'irritable people' are more likely to develop heart disease and high blood pressure levels than 'non-irritable people' (Smith & Fernhall, 2022). When a person is angry, catechol secreted by the body will weaken the immune system. Anger is not only physically harmful to the person who experiences it but also weakens their ability to forgive. An angry person will have a germanium sensation and feel full of strength and courage. Therefore, when a person is angry, a common reaction is to make a noise or fight (Markussen, 2017; Tashjian et al., 2022). Several of the participants stated that they primarily vent their frustrations to their loved ones. They felt that the interpersonal tension brought on by their work made them unbearable and hoped to leave their current design teams so that they could breathe a sigh of relief. Some participants had resigned from design projects to resolve problems related to 'anger'. However, this solution only had a short-term effect and addressed the symptoms rather than the root cause of the problems. When their interpersonal relationships in a new work environment were as tense as in the environment they had left, the participants repeated the same mistakes and 'anger' again affected their personal lives. The participants discussed alternative approaches. One participant suggested that emotions should be vented, while others suggested they should be suppressed. EI dictates that neither venting nor suppressing emotions is optimal. The best approach is not to get angry often. This approach requires one to control the negative emotions before getting angry and before the body enters a state of tension.

6.2.2.2 Emotional Elements That Will Enhance Junior Design Students' Abilities to Manipulate Their Design Processes

Most participants were unaware that the eight fundamental emotions are categorised according to consumers' reactions during product use as per Herrmann et al. (2018). Therefore, the data suggest that these emotions can be caused by an individual's external environment; however, they cannot directly cause cognitive processes. Thus, this study revealed a void in prior design and emotion research. Designers' emotions, while responding to their external environment, and thus their decision-making, do so differently than users' emotional changes which reflect responses to their external environment, with users' emotional responses usually being captured via consumer feedback forms or similar; however, these emotional responses usually do not create further design choices for designers.

Therefore, it is necessary to identify and analyse functional emotions influencing the design process itself. Upon conducting reviews of psychology studies, the research team identified basic emotions and examined the theoretical explanations of each within existing design literature. The studies reviewed provided identification of many common types of emotions individuals experience during the design process. These types of emotions were considered and a new conceptual model (see Figure 2.1) was developed to illustrate how emotions relate to those that occur most frequently during the design process.

---

1. **Sadness**
   The state of being distressed and sad to the point of crying. It is usually accompanied by feelings of weakness, frustration, disappointment or fatigue and physical reactions such as chest tightness, a stiff expression, decreased willingness, low mobility and crying (Shirai et al., 2021).
2. **Nervousness**
   The state of feeling nervous when facing a situation or something you are not used to. It is usually accompanied by physical reactions such as stiffness, the contraction of blood vessels or muscles, an increased heart rate, sweating, loss of appetite and difficulty breathing (Lewis, 1995; Turner, 2021).
3. **Hatred**
   Feeling dissatisfied with another party's bad attitude or bad treatment of others and having a strong desire for revenge. The stronger the feeling of hatred, the greater the inability to calmly judge emotions (Rasouli et al., 2022).
4. **Regret**
   After an incident, people sometimes regret what they did or did not do. Continuing to feel regret may adversely affect judgement or performance (Li et al., 2021).

5. **Frustration**
This is the feeling that what is expected will not come true, a feeling of hopelessness. Sometimes it can yield a positive evaluation ('thinking about it') and other times a negative evaluation ('if you give up, it will be over'). This is a mixed emotion (O'Brien et al., 2022). Frustration is often accompanied by a lack of motivation to do what you want to move forward. Although 'lack of vitality' and 'lack of motivation' are often considered to be the same, they are completely different states, and the accompanying effects are different (Bahlgerdi et al., 2021). Frustration often brings the feeling that all is in vain, accompanied by an inability to gain control'. When people fall short of achieving a goal or give up on a goal, they often experience a sense of emptiness (Bares et al. 2021).

6. **Discouragement**
A feeling of disappointment when things do not develop as expected. When an outcome yields unsatisfactory results, people will become discouraged when they realise that the outcome is not the fault of others but the result of shortcomings related to their own imagination and analysis (Uzun & Yıldırım, 2018).

7. **Loneliness**
The feeling of lacking something you want or should have; feeling uneasy or unfulfilled or feeling desolate when you wish to see someone. Depending on the degree of loneliness, physical reactions such as chest tightness and aimless wandering may occur (Solmi et al., 2020).

8. **Disgust**
Feeling an aversion towards unpleasant people or things. The degree of disgust experienced may cause different physical reactions, such as goosebumps, chills or nausea (Briñol et al., 2018; Zanganeh & Hariri, 2018).

9. **Inferiority**
An 'inferiority complex' refers to the feeling that one is somehow 'less' than others. It arises primarily from comparison to or competition with others and can give rise to emotions such as melancholy or jealousy (Roseman et al., 2021; Tracy & Robins, 2008).This also describes an individual's belief that another person's personality or abilities are inferior. The individual judges the other's behaviour and considers it to violate social rules and, as a result, despises them (Park & Baek, 2018).

10. **Anger**
When you feel that you have been attacked physically, mentally or socially, you may lose your composure. This emotion is typically accompanied by changes in facial expressions (frowning, lifting the corners of the eyes and turning the corners of the mouth downwards) as well as physical reactions (increased blood pressure, faster heartbeat and trembling) (Briñol et al., 2018).

11. **Anxiety**
A feeling that someone or something you value is threatened or that something bad is going to happen and being unable to calm down. It typically causes physical reactions such as chest problems and stiff expressions (Reinholt et al., 2022).

12. **Jealousy**
A feeling of hate when someone you love falls in love with someone else or when someone is better or richer at something than you are. Jealousy can be described as a combination of various emotions, such as anger, anxiety, hatred, powerlessness and low self-esteem (Dawood & Kashif, 2021).

13. **Dissatisfaction**

An inability to accept the status quo. The emotions that arise when desires are pursued or when one's abilities or the surrounding environment are considered insufficient are obvious (; Yim & Byon, 2018).

14. **Fear**

The state of being afraid of being physically or mentally coerced or being injured by things beyond one's power. It usually causes physical reactions, such as chills, rapid heartbeat, tremors and cold sweats (Young, 2019).

15. **Joy**

A feeling of happiness when something good happens. When one's mood is elevated, depending on the degree of joy, there may be physical reactions such as laughing, increased activity, improved athletic ability, rapid heartbeat and crying (Kumar et al., 2019; Zhou et al., 2021).

16. **Happiness**

A feeling of satisfaction and elation. Similar to 'joy', depending on the level of the emotion, there will be physical reactions such as laughing, increased activity, improved athletic ability, rapid heartbeat and crying (Naseem, 2018).

17. **Pleasure**

When an expectation is fulfilled, the soul is satisfied and one's mood is one of extreme satisfaction. If the level of happiness is very high, it may cause physical reactions such as warmth in the chest and crying (Li et al., 2020).

18. **Longing**

A desire to be with other people who are in the same situation or are experiencing the same mental state as oneself; to experience a sense of closeness (Jiang & Koo, 2020).

19. **Affinity**

An attraction to ideal people or things; a feeling of yearning or wanting to target such individuals (Kim & Park, 2013).

20. **Curiosity**

Having a strong interest in the unknown and a desire to explore the nature of things. Each person has a different degree of 'curiosity', depending on the amount of dopamine their brain secretes (Buheji, 2019).

21. **Calmness**

A mood that is neither elevated nor panicky but relaxed. When a person is in a calm state, reason supports their actions and correct judgements will be made (Pawar et al., 2018).

22. **Pride**

A focus on honour (praise and appreciation for one's abilities or behaviour or the desire to be honoured (Rendina et al., 2019).

23. **Excitement**

Emotions are high and stimulated, and the mental state is unstable. If the level of excitement is strong it may cause physical reactions such as increased blood pressure, a faster heartbeat, tremors, faster and shallower breathing and crying (Tracey & Hutchinson, 2019).

24. **Anticipation**

Preoccupation with the future, hoping and waiting for something to happen. If the level of anticipation is high, it may cause physical reactions such as an elevated mood and increased heartbeat (Christner et al., 2020; Rathore & Ilavarasan, 2020).

25. **Superiority**
Emotions that affirm oneself as better than others in a competitive situation or a feeling that one is better than others in general (Mennin et al., 2018).

26. **Sympathy**To take on the grief, pain and misfortune of others as one's own; caring about how others feel. If the feeling of sympathy is strong, you may be like the other person. This state typically causes physical reactions such as chest tightness, crying and elevated emotions (Uzun & Yıldırım, 2018).

27. **Respect**
A feeling of dignity is experienced when others admire your personality, knowledge, experience, etc. Respect means recognising and admiring the praiseworthy behaviours and achievements of others (Brom et al., 2018).

28. **Courage**
A positive and resilient attitude towards things that cause others to feel fear, anxiety and hesitation (Howard, 2019).

29. **Affection**
A feeling of love for people or objects. 'Cherish one's offspring' is the innate impulse that forms the spine of compassion
(Kumar et al., 2019; Schrage et al., 2020).

30. **Healing**The pattern of becoming sick or experiencing sadness and getting better. The healing mood is deeply related to brain chemicals such as serotonin and oxytocin (Uwajeh et al., 2019).

31. **Gratefulness**
Gratefulness refers to the state of being thankful, appreciative and acknowledging the goodness, kindness or positive things that one receives or experiences. It involves recognising and expressing gratitude for the people, situations or blessings in one's life. Gratefulness encompasses a deep sense of acknowledgement and appreciation for the positive aspects of life, which can include tangible things, experiences, relationships or personal qualities
(Achmad et al., 2023; White et al., 2023).

32. **Relief**
The feeling that one has no worries and where anxiety has been replaced by peace of mind (Harley et al., 2019; Naismith & Lajoie, 2018).

33. **Touched**
A feeling that arises from contact with beautiful and exceptional things. It causes a strong impression or shock, and then one becomes fascinated or moved by what has been experienced. When this emotion is strong, it may cause physical reactions such as emotional upswings, anxiety and crying
(Casals-Gutiérrez & Abbey, 2020; Chen et al., 2018).

34. **Energetic**
The human brain's greater energetic output is connected to its superior cognitive capacities and the broad variety of actions exhibited by humans.
This expense is a result of the evolution of the human brain
(Bordone et al., 2019; Ciompi & Panksepp, 2005).

35. **Enjoyment**
Humans are motivated by enjoyment, which drives designers to accomplish activities that are generally beneficial to designers. It stimulates designers to participate in activities that are vital to their role in structuring human lives to optimise their enjoyment
(Bastian, 2017).

36. **Satisfaction**Being satisfied is defined as a state of well-being characterised by low arousal. When we are satisfied with something, we are linked to it. This creates a sense of completeness within us and, as a result, we become tranquil and can experience enjoyment in the moment (Liu & Sourina, 2012; Yim & Byon, 2018).

37. **Peacefulness**

    Peacefulness refers to a mental state of tranquillity or sereneness. It may indicate the absence of concern and anxiety. When designers' minds are cluttered by worries and task lists, it can be beneficial to replace thoughts with more peaceful ones (Ren et al., 2026).

38. **Stressed**

    Stress is a natural response to the daily strains of life. Worry, fear, rage, grief and a variety of other emotions are all natural reactions and necessary components of existence. However, if the tension that lurks behind these feelings becomes a hindrance to a designer's capacity to perform required tasks, stress becomes hazardous (Qin et al., 2020).

39. **Commitment**

    Commitment has been found to act as a moderator between designers' negative emotions and supportive actions. Furthermore, it has a considerable influence on the shaping of design behaviours. It has been shown that certain design issues are associated with negative emotions rather than good emotional reactions throughout the creative design process. These findings have implications for reducing the stress designers experience, increasing commitment to design optimisation and securing support for sustainable design development (Lemay Jr & Dobush, 2015).

40. **Interest**

    Interest refers to the emotion of wanting to pay attention to something or wanting to feel connected with and know more about something. The evaluation viewpoint is a useful approach for constructing underlying motivations (Silvia, 2005).

Types of Emotion Identified in the Design Process of Junior Design Students

Junior Design Students identified 40 different emotions in their Phase 1 and 2 design project processes through a series of focus groups. The amount of time students worked during the two phases was similar.

Emotions that appeared most often in the junior design student's discussion included feelings of being calm, which accounted for 5.33% of the total number of references to emotion; anxiety, being moved, commitment, healing, curiosity, excitement and enjoyment, which comprised 4.00% of the references; and other emotions including hatred, regret, discouragement, fear, longing, sympathy, peacefulness, interest, pleasure, joy, happiness, courage, relief and having lots of energy, which made up 2.67% of the references. Emotions that were least referenced, or at 1.33% of the references, were anger, jealousy, sadness, frustration, loneliness, disgust, inferiority, stress, affection, satisfaction, affinity, pride, anticipation, respect, gratitude and superiority. These are summarised in Table 6.1 and include the number of times they occurred and their percentage of references to emotion.

TABLE 6.1
The emotions and their occurrence noted in participants' transcripts

Emotion category	Emotions experienced in the design process	Frequency of the emotions mentioned in the participants' transcripts	Percentage of the emotions mentioned in the participants' transcripts (%)
External negative	Nervousness	3	4.00
	Hatred	2	2.67
	Regret	2	2.67
	Discouragement	2	2.67
	Anger	1	1.33
	Jealousy	1	4.00
	Dissatisfaction	3	2.67
	Fear	2	2.67
Internal negative	Sadness	1	1.33
	Frustration	1	1.33
	Loneliness	1	1.33
	Disgust	1	1.33
	Inferiority	1	1.33
	Anxiety	3	4.00
	Stressed	1	1.00
	Longing	2	2.67
Natural	Calmness	4	5.33
	Sympathy	2	2.67
	Affection	1	1.33
	Touched	3	4.00
	Satisfied	1	1.33
	Peacefulness	2	2.67
	Commitment	3	4.00
	Interest	2	2.67
Internal positive	Pleasure	2	2.67
	Affinity	1	1.33
	Curiosity	3	4.00
	Pride	1	1.00
	Anticipation	1	1.33
	Respect	1	1.33
	Healing	3	4.00
	Gratefulness	1	1.33
External positive	Joy	2	2.67
	Happiness	2	2.67
	Excitement	3	4.00
	Superiority	1	1.00
	Courage	2	2.67
	Relief	2	2.67
	Energetic	2	2.67
	Enjoyment	3	4.00
	Total	75	100.00

6.2.2.3 Principles for Guiding Junior Design Students to Adjust Their Emotional Changes in the Design Process

This research study presents a series of recommendations or core ideas aimed at assisting junior design students in regulating and responding to their emotional changes during the design process. It implies that there are particular techniques, tactics or guidelines that may be used to assist young design students in navigating the emotional ups and downs that are common during the design process. This might include ways for dealing with dissatisfaction, handling creative blockages, keeping motivated and dealing with the uncertainties and obstacles of design work. The principles may also include strategies for increasing self-awareness, resilience and a healthy mentality, ensuring that students are prepared to deal with the emotional components of design as they build their talents and professional identity (Mennin et al., 2018). This is comparable to a situation where an obese individual must first acknowledge that they cannot keep increasing their weight, prompting them to regulate their food intake.

If they cannot help but eat more than a standard meal, they can adjust the amount of food consumed at their next meal. The principle for controlling anger is similar to that in this example.

Research on EI proposes the following three specific methods to control anger, and the feedback from the participants reflects similar sequential and interrelated steps.

(1) The potential physical harm of emotions

It is vital for individuals to comprehend that harbouring anger might result in an untimely demise. Once a person accepts this concept as true, they will be willing to change themself. As this notion becomes a conscious thought, a person can stop themselves from easily becoming angry. Being hostile towards others can make a person angry. Accordingly, the thought that hostility can endanger one's own life should help a person avoid being hostile towards others.

(2) Recognising occasions that give rise to emotions

Once a person knows that anger can be physically harmful, they need to be able to recognise what events will easily arouse their anger. Once they know the triggers, they will be able to stop themselves from becoming angry. For example, if a person knows that meeting their sister makes them irritable, they immediately become irritable, and every time they pass through the accounting department, they speed up their pace so as not to cause themselves physical harm.

(3) Learn how to regulate negative emotions

When the person applies the above two steps and still becomes angry, they must learn how to regulate negative emotions and calm themselves down. The participants shared that when they realised that they were angry, they would immediately remember their favourite Korean pop music and replay it in their minds. Everyone deals with negative emotions differently. Accordingly, they must understand what can make their upset state return to normal.

Two additional steps proposed in the earlier chapters were supported by the participants' feedback. However, these were not investigated in the previous EI proposal research studies.

6.3 RESEARCH STUDY 2

This study aimed to determine whether emotional changes are involved in the design process. Although some scholars have suggested they are, little evidence has been provided to support this. This study supports the notion that emotional changes influence conscious experiences and decision-making during design processes.

6.3.1 Phase 1: Investigation of the Emotional Elements in the Design Process

6.3.1.1 Self-Recognition of Emotions Using the eMoods Application

The participants tracked their emotions using eMoods during Research Study 2 via self-recognition of emotional changes. Using the eMoods application, the participants were able to identify their feelings – 16% of participants identified anxiety, while another detected irritability; 32% of participants noted interest; 32% of participants identified pleasure. No participants indicated their emotions as depressed, sad, elevated or uncategorised emotions. These findings show that participants could identify basic emotions. The participants had a preliminary awareness of their emotions during the design process. These findings additionally show that positive and negative emotions are involved in the design process.

6.3.1.2 Emotion Detection and Tracking in the Design Process

Prominent Emotional Changes among the Participants during the Design Process

Scientific records on emotional changes were obtained from this research study. A systematic review was conducted using the eMoods application to identify prominent changes in the participants' moods during the design process. Participants were able to record potential problems concerning design practices by using the mobile application. Previous studies that employed mobile applications to monitor mood fluctuations among users have demonstrated the eMoods application's efficacy (Caldeira et al., 2017). Scholars have praised the application's functions for documenting, reporting and graphing emotional changes. This mobile application was used in this research to conduct a thorough study and identify junior design students' optimal emotional reactions that can improve their decision-making abilities throughout the design process.

6.3.2 Phase 2: Focus Groups

6.3.2.1 Facial Expressions Detected by CAPTIV Reveal Emotional Changes during the Design Process

Several obvious facial expressions were detected by CAPTIV, but the participants could not accurately present any emotional changes as a continuous set. Participants could identify the emotions they experienced with guidance from the eMoods application. However, emotions are continuous experiences that require a comprehensive approach in terms of observation. The participants could not recognise detailed emotional changes during the design process. They failed to understand that emotional responses typically entail more than a single, isolated identification and primarily manifest as a series of ongoing changes. Participants successfully discerned distinct emotional changes linked to the use of the programme. Emotion is an ongoing response that may be discerned using a pragmatic approach. However, the individuals had difficulty in discerning subtle variations in emotion between different conditions. During the design process, participants did not grasp the fact that emotional responses were not identified as a single occurrence, but rather as a sequence of emotional alterations.

The CAPTIV L7000-Premier application was used to record participants' facial expressions during Phase 3 of the Research Study 2. The unique advantages of the facial expression tracking system's ability to manipulate facial expression detection and identification were explored (Jiang et al., 2022) and showed that the participants were not familiar with the use of emotions when designing. They were not sensitive to recognising the emotional changes they experienced during the design process (Figure 6.2).

6.3.2.2 Emotion Management in the Design Process

Emotion Adjustment Speed May Affect How Designers Perceive Emotional Stimuli

Emotion adjustment speeds may influence how designers perceive and manage emotional stimuli (Figure 6.3). Reflected by the statistical data collected during Phase 3 of the Research Study 2, the speed at which decision-making took place was higher for participants who completed the design phase (with response times between 5.12 and 3.03 seconds and an average decision-making time of 14.05 minutes) than for participants who did not (with response times between 3.15 and 2.05 seconds and an average decision-making time of 27.05 minutes). In addition, recent research indicates that negative feelings impact how we view scheduling and resource allocation risks (Holmes et al., 2019; Khan et al., 2020; Vagnoni et al., 2012).

6.3.2.3 Emotional Shifts May Impact Designers' Perceptions, Decision-Making and Design Manipulation

The results of this research are consistent with previous findings and can be expanded by indicating that the degree of emotional changes may influence designers' perceptions, decision-making and manipulation of the design process. Inefficient

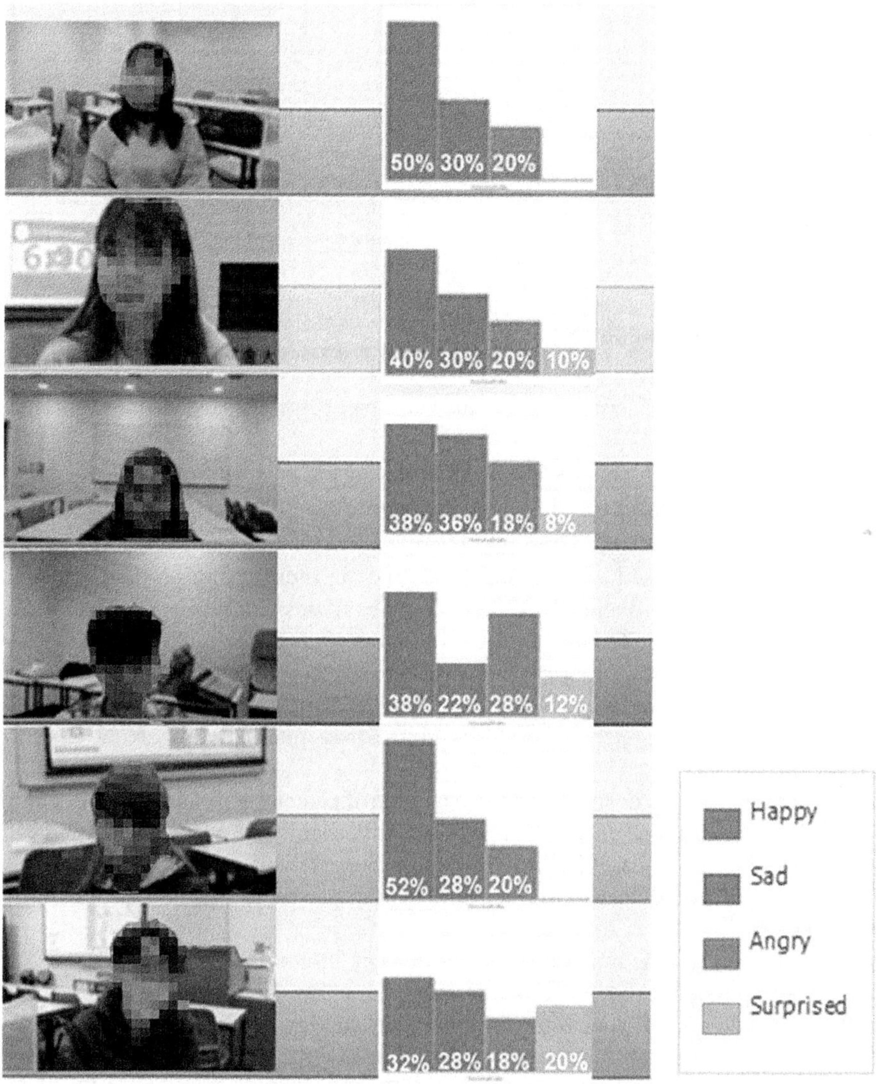

FIGURE 6.2 Using the CAPTIV L7000-Premier application to record participants' facial expressions

decision-making and design process difficulties increased when the participants experienced relatively negative emotions, such as anger and sadness. When participants felt angry, their ability to carry out design solutions decreased, as they were exposed to highly arousing cues that indicated great concern.

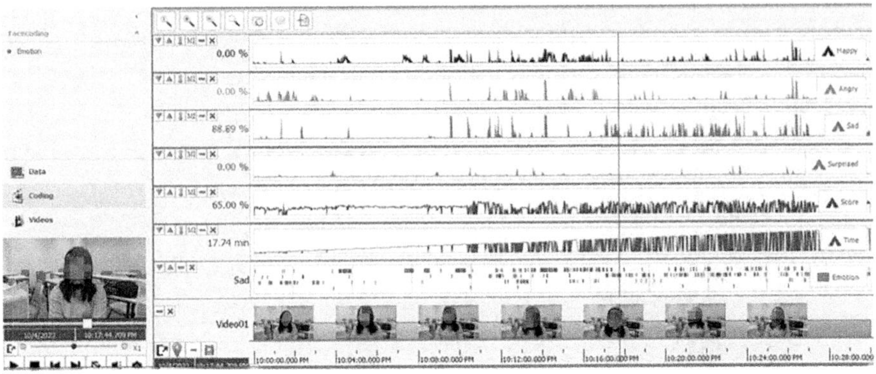

FIGURE 6.3 Emotion adjustment speed may influence perception and emotional stimulus

6.3.2.4 Emotional Changes Were Modulated for an Optimised Design Process and Design Outcome

Participants modulated the emotional changes evoked when engaging in different design aspects and based on personal experience in their design process. The evaluation of tension is influenced by designers' awareness and management of their emotional fluctuations. This helps them maintain a consistent estimate of the level of positive emotion they experience during the design process (Streiner & Norman, 2011). Both positive and negative stimuli have a comparable impact on emotional experience under the supervision of emotion management (Baumgartner & Pirker, 2019).

An analysis of the design process based on the facial expressions collected in Phase 3 of the Research Study 2 provided further intriguing findings. Emotional tracking had an obvious effect on the design process and outcomes. Accordingly, the data are presented as a source for new hypotheses rather than as a confirmation (Figure 6.4).

With the use of new technologies and devices in emotion sensors (CAPTIV), it has become possible to identify and measure emotions. These technologies provide relatively accurate and clear information about emotional changes and can be used to highlight the importance of emotional support as an essential concern for designers. The adoption of an emotion detection system can help researchers and designers understand the emotional changes of users and designers. The data can convince them to introduce emotional concerns into their design process.

The statement asserts that the presence of numerous superfluous emotional cues from both the visual and auditory sensory channels enhances comprehension of an individual's emotional condition as conveyed through experiences including both visual and audio elements (bi-modal events) (Oh et al., 2011). This suggests that the presence of both visual and auditory clues, which are consistent throughout, enhances one's ability to accurately assess the emotional aspect of an event. The statement suggests that this issue is well acknowledged and embraced in study and scientific comprehension. Put simply, when visual and auditory cues are present and

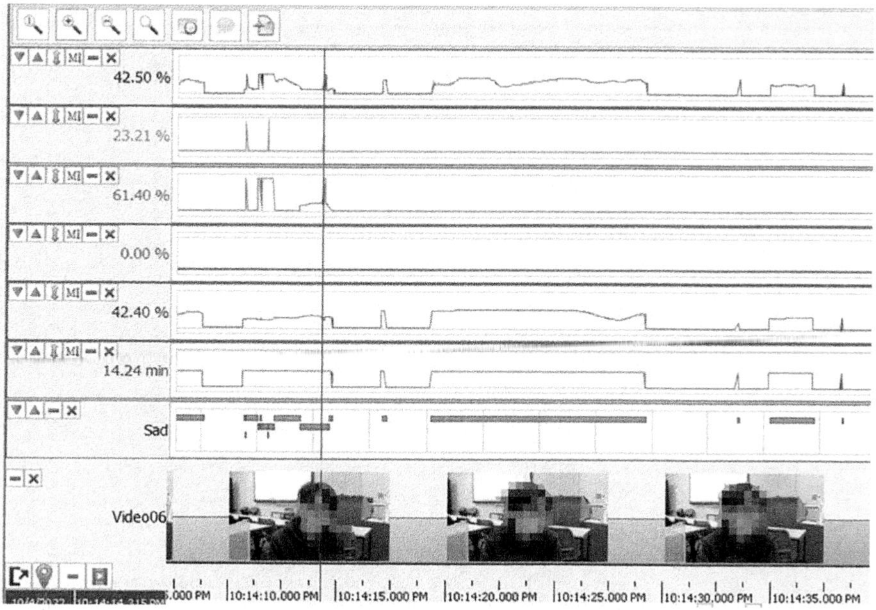

FIGURE 6.4 Emotional tracking had an obvious effect on the design process and outcomes

synchronised in conveying a certain emotion, individuals are more able to understand and perceive the emotional state being sent (Pawar et al., 2018). This pertains to the hypothesis that several sources of information might augment and increase our understanding of emotions conveyed in bi-modal events. Using categorisation tests in which the participants had to choose between two or more affective states (with one matching the affective state elicited by a stimulus), it was found that integrating auditory and visual cues has an impact on determining the emotional value of events (Qin et al., 2020). The background theory of emotion employed a psychophysiological framework for evaluating emotions. Additional emotion assessment methods were developed by obtaining objective data on designers' emotional changes. The role of emotion in the design process through the scientific facial expression sensor CAPTIV was examined. The approach towards managing emotions during the design process was demonstrated. As a result of using quantitative measurements and electrodermal activity measuring tools, the role of emotions were be revealed.

6.3.3 Phase 3: Individual Interviews

6.3.3.1 Individual Interviews and the Reflective Reviews in Lablog

In this stage, the three highest and three lowest academic achievements among the participants were invited to attend a 20-minute interview as well as a digital lablog review. The objective of this phase was to understand their emotional changes and the influence of emotions on their design processes. This phase was audio-recorded

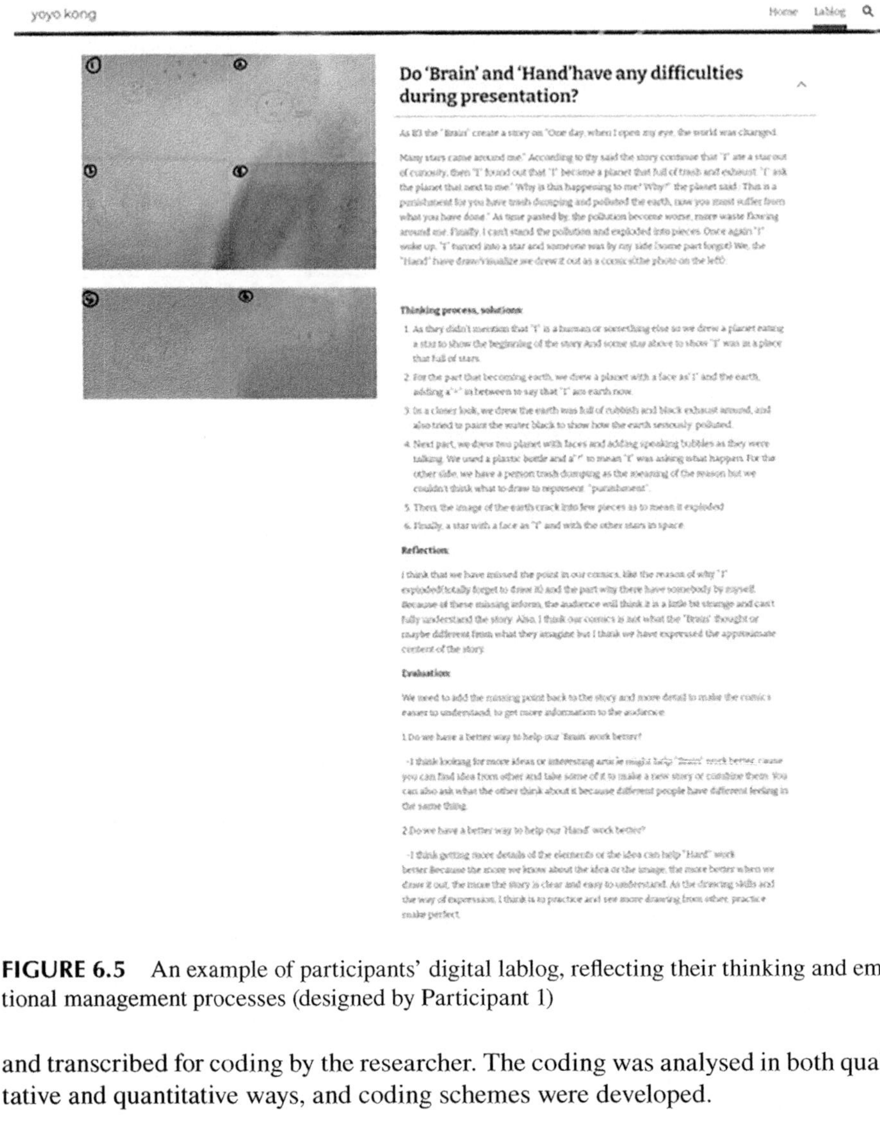

FIGURE 6.5 An example of participants' digital lablog, reflecting their thinking and emotional management processes (designed by Participant 1)

and transcribed for coding by the researcher. The coding was analysed in both qualitative and quantitative ways, and coding schemes were developed.

Reflective Review in Participants' Lablog

The lablog served as a significant record of the learning process and worked well as a space in which to try new things, a reflective journal and a thinking tool (Figures 6.5 and 6.6).

Since each participant's digital lablog provided a detailed account of the order and duration of the aspects they focused on during the design process, it allowed them to use their time effectively. On reflection, this could help them establish effective time management for future projects.

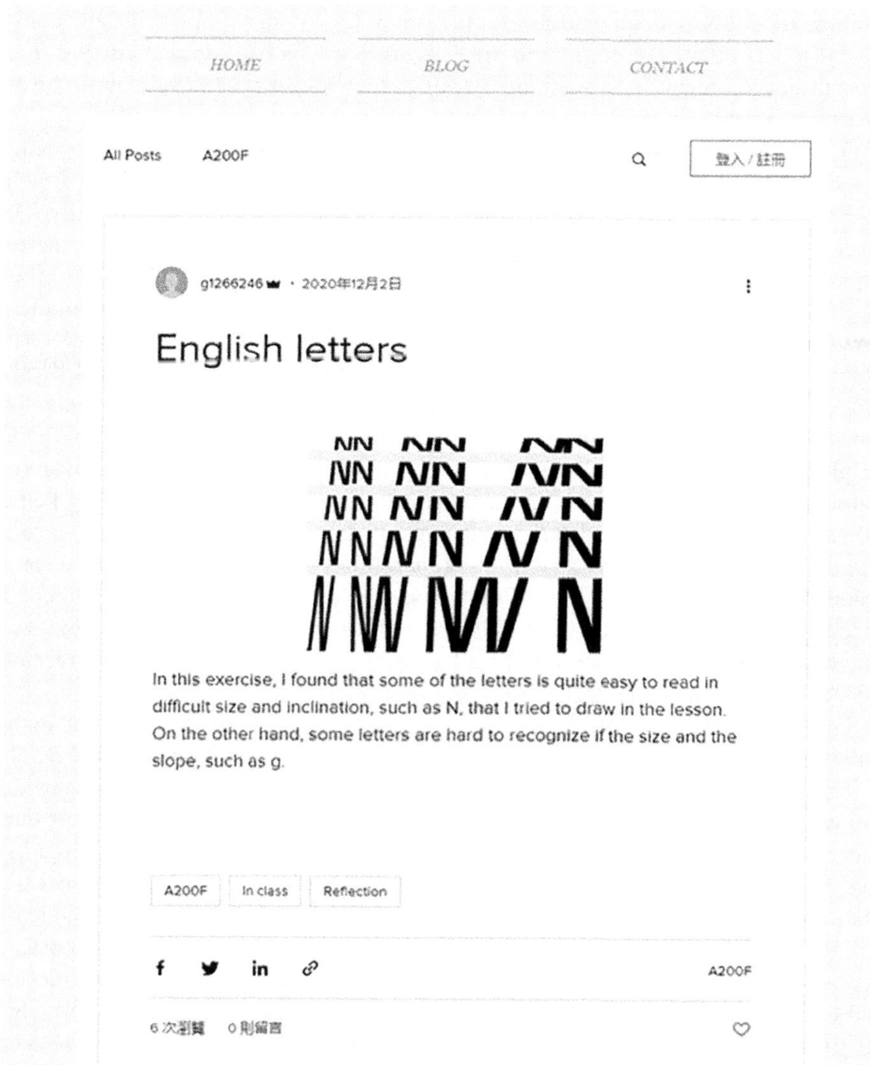

FIGURE 6.6 An example of participants' digital lablog, reflecting their thinking and emotional management processes (designed by Participant 2)

As a tool with which to try new things, the lablog provides a notebook for the recording of dates and places, people encountered, observations of phenomena, new thoughts and feelings, experiences and knowledge and journeys of discovery. As a reflective journal (Murty & Purcell, 2007), it provides an invaluable database for revising and improving students' thinking and problem-solving processes. It is a record of thoughts, impressions, opinions, action strategies, perceived outcomes and results. As such, it serves to present new possibilities and expand opportunities. As a thinking tool, it provides a platform for the generation of ideas, a portable and convenient place in which to record mind maps, brainstorming lists, answers to questions, concept development and design analysis through revisiting and re-examining the creative process.

For this study, lablog was considered to be the best tool for conditional knowledge, as it could be used to record all the information about the learning process. It helps students learn why and how to apply knowledge declaratively and procedurally. This can help students make strategic decisions regarding resource allocation, thereby improving the effectiveness of their strategies and design solutions. The thinking record ('lablog') of Leonardo da Vinci is regarded as the most typical example of how a lablog can help learners process conditional knowledge. The adoption of the lablog eventually encouraged the participants to evaluate how the proposed principles could assist them in executing the design process using metacognition. The concept of metacognition can support students in the design research field to provide insightful feedback. The process of planning, applying and evaluating (all thinking activities and problem-solving and evaluation procedures) using the lablog helped the participants enhance their knowledge and awareness and regulate their learning progress through evaluation. From the preparation of the presentation on metacognition, insights into leading metacognitive strategies in learning were obtained. To conclude, the lablog served as the ultimate tool for tracking participants' emotional changes and discovering answers to all possible questions related to the research subject.

The knowledge obtained by participants as a result of gaining a better understanding of design aspects can be used for different tasks. Participants will be able to implement effective approaches for communicating with others when engaged in teamwork, particularly if they work with others who similarly improved their interpersonal skills. This learning experience introduced the application of metacognitive control and self-regulatory processes to the participants' design learning processes. They applied cognitive processes to control and regulate their emotions while learning, including planning, monitoring and evaluating. The knowledge and experience obtained from this learning activity (e.g., better understanding of their emotions and user experiences) included awareness and reflection on investigative methods to obtain crucial content, that is, deep knowledge versus partial knowledge, which can lead to improved design outcomes. The objectives of this stage were to discover the influence of emotional changes on participants' design processes and participants' knowledge about emotion in design and to investigate the effectiveness of collectively applying the proposed principles to the knowledge of design and emotion, and the elements involved in the principles that motivate junior design students to learn.

6.3.3.2 How to Motivate Junior Design Students to Manipulate Their Design Process Using Managed Emotional Elements

Although there are still many unclear points about how emotions are triggered, the operating mechanism is as follows.

To understand the mechanism that triggers emotions, one must know the 'belief system'. A person's worldview comprises memories stored in the temporal lobe and their belief system (a combination of various perceptions) in the prefrontal lobe. The so-called belief system refers to 'I am such a person', 'how the world works' and similar concepts. Everyone has a fixed mindset (belief) and is used to judging various things. Beliefs include not only socially positive thoughts but also discriminatory ideas and self-centred ways of thinking. Beliefs are like a map of the world that includes oneself and a kind of inner expression. As it is deeply affected by memory, the map refers not only to guiding behaviour but also to the 'self' itself.

In terms of the brain processing information, emotions are closely linked to the amygdala and hippocampus of the brain's limbic system (Nunez & Srinivasan, 2006). The hippocampal gyrus can store short-term memories for up to days and plays the role of a door to the long-term memories that are stored in the temporal lobe. The hippocampal gyrus also compares externally stored information with old memories. The amygdala interferes with the hippocampal gyrus, which can enhance or weaken an original memory, and is associated with all parts of the brain, acting as a centre that processes emotions.

When external information enters the brain through the sensory organs, it will be compared with memories that are stored in the hippocampus and the temporal lobe. The information is not necessarily derived from the physical environment or other sources, such as movies, novels or the imagination (as long as the setting resembles reality) (O'Brien et al., 2022). This allows hypnotists to make people feel pain and sadness without there being a physical reason. The opportunity may come from the outside world, but all information is generated in the brain. Once the information has been compared with the memory in the temporal lobe, it is divided into positive or negative events. The information is then evaluated in cooperation with the protomala, ventral tegmental area, insula, nucleus accumbens and the prefrontal lobe, which houses the belief system. This leads us to generate all kinds of emotions. The more significant the difference between the information and the self-belief system, the stronger the emotions will be, and the amygdala will enhance the degree of emotional performance.

Several brain substances are released during this process. When we are sad, norepinephrine is released; when we are happy, dopamine and serotonin are released. The mechanism is the same. It then transmits emotional information through the hypothalamus, which, in turn, affects the body's response. Norepinephrine is a neurotransmitter in the brain. When it enters the body as an endocrine substance, it will cause a reaction that stiffens the face or accelerates the heartbeat due to fear. Regardless of what brain substance is secreted, serotonin will eventually be released. Therefore, no matter how terrifying an experience, how intense an experience of sadness or how intense pain may be, a person will eventually become calm again. Although the above process is presented in a simplified manner, the mechanism that triggers emotions is approximately the same. However, as mentioned in the introduction, the mechanism itself is not particularly important. It is essential to understand the 'texture' of emotions, to treat emotions as entertainment and to control them to achieve goals.

6.3.3.3 The Key to Mastering Sadness: The Frontal Lobe

We experience sadness in various forms. Sadness is closely related to the prefrontal lobe, which processes abstract information. The state of being distressed and sad to the point of crying is typically accompanied by weakness, frustration, disappointment and fatigue, and it often causes physical reactions such as chest tightness, a stiff expression, reduced willingness, low mobility, being locked in one's own world and crying. Even cats and dogs, with their underdeveloped frontal lobes, will feel sad. However, only people experience sadness about events that are not happening in the moment, such as remembering sad events that occurred in the past.

According to the feedback obtained from the participants, they had not previously learned any concepts related to emotion management and self-awareness. Some participants were unsure about the specific emotions they experienced while working on the assigned design task.

The participants noted that designers will be able to work well if they are calm because they will be able to think logically and make good decisions. Participant 5 noted,

> *'I think a calmer and less emotional state can improve my learning because I can [take on] more knowledge when I am calm'.*

The feedback from participants indicated that when they felt calm, they put aside their feelings and problems and thought about the design problems objectively. This finding enhances the knowledge about the function of emotion in the design process, as noted in the literature review. Dopamine and serotonin are secreted when the brain is in a calm state (Horvath et al., 2018). An active prefrontal brain allows designers to think rationally and make sound decisions. Participant 5 noted,

> *'I think I feel better when a person settles down or is still, because they will concentrate. I think a peaceful and calm environment is better because if there are too many people talking on the side, I can't concentrate on the design, but when I'm alone, I can go and think clearly about it – what an idea'.*

Designers have to consider a wide range of situations, including long-term concerns about the relationship between themselves and target customers. Calm people will be able to make good decisions because they will have peace of mind. Participants' feedback confirmed Liang and Jin's (2011) findings about the influence of calmness on designers. If designers can work calmly, they will be able to respond to problems quickly without relinquishing logical thinking.

Participants emphasised that both negative and positive emotions can help designers work effectively by contributing different aspects. People experience a wide range of emotions, as noted in the literature review (Desmet, 2018). An experience is more satisfying when it is not dominated by any one emotion. Positive emotions can help designers communicate well with others. Participant 6 stated,

'I think it may be more positive to be able to communicate with other people about emotions, and it will be easier to learn if you communicate or talk with other people more'.

Participant 4 stated,

'I feel happy and excited. Listening to my own preference will make me comfortable, and I think anxiety will be effective for me because when I'm stressed, I am anxious and want to get things done quickly'.

Alternatively, some participants argued that designers should be able to avoid threats by relying on knowledge and theory rather than on fear-based information. For example, designers can prevent risk when they learn to cope with possible threats while selecting raw materials. The participants agreed that designers can comprehend abstract and potentially dangerous concerns. Participant 1 noted,

'I would feel the difference between me and Cherry. In my case, personal comfort is the priority. Cherry is able to cope with life first. It seems that I have been forced out of some ideas, so I prefer to go there and find a point that interests me, and then I will go deeper into it and do my homework'.

Participants' feedback confirmed that emotions have the potential to influence one's perception of danger through a fundamental mechanism inherent in threat detection. Designers were instructed to make rapid judgements about whether persons were carrying firearms or neutral items while experiencing various emotions (Ho & Siu, 2012). Anger increased the likelihood that neutral aspects would be mistaken for those associated with violence, but not vice versa. Negative emotion-induced bias in design conceptualisation can be corrected with the appropriate skills.

Some participants emphasised that they worked better when they applied curiosity. Participant 1 noted,

'Personally, I think it is more effective to be curious and interested and achieve my expected learning outcomes ... Because only when I am interested in ... things will I ... By fostering a sense of motivation and anticipating the outcomes, I can bolster my resolve, enabling me to approach my design assignments with renewed vigour. This mindset encourages me to persist in my efforts, continuously refining my homework designs and gathering a diverse array of materials to support my creative process. [If] you are interested in doing [a task], it will not be so hard'.

Participant 3 agreed:

'I think when I learn something that I find interesting and when I talk about it in class, I am interested in it and passionate about a [specific] design ... In other words, you will push yourself to learn, that is, you will make

yourself want to learn more, and if you have homework, it will help you learn'.

Participant 4 stated,

'I think [elevated] emotions will help me learn a lot because excitement improves my mental state. This kind of emotion will make me learn faster'.

Some participants noted that curiosity made them practice more. Participant 2 stated,

'[If] you're not interested in [something], you will be unable to achieve your expected learning outcomes'.

The feedback from participants indicates that the concept of curiosity plays a role in the design process, which is in line with the findings of previous studies. Curiosity is one of the emotions that often results in information-seeking and motivation. A design using curiosity principles can adequately arouse the interest of participating designers to engage with the design.

6.4 CASE STUDIES

6.4.1 Case Study 1: Headspace Meditation App (Emotional Design for Stress Reduction and Habit Formation)

Headspace is a well recognised emotion-driven digital design programme, renowned for its meditation and mindfulness features, with over 100 million downloads internationally (Jablonsky, 2022). The software features a simple design aimed at fostering a sense of tranquillity and alleviating stress, facilitating behavioural change by converting significant psychological ideas on emotion regulation into practical, user-friendly tasks. Headspace offers insights on how visual choices, interactions, and user mapping correlate with user emotions through successful case studies, mainly aimed at educators and younger designers to enhance engagement and well-being.

The Headspace emotional design framework is founded on three fundamental ideas. Eliminate all recollections; they were deliberately founded on scientific study about colour psychology and visual hierarchy. To enhance the soothing blues and greens, the application incorporates brighter oranges and yellows – warm hues that convey optimism (and sophistication) – to achieve content equilibrium. This dual colouring method enabled us to represent the 'dual emotional needs' throughout design phases, such as providing customers with energy during onboarding and inducing relaxation during meditation. The interface eliminates high-contrast or dense features that studies indicate might elevate stress levels, particularly among novice users. Secondly, Headspace utilises micro-interactions and incremental disclosure, shown by an animated breathing exercise and expanding rings that provide real-time biofeedback. These animations are categorised as 'transitional prods', designed to assist users in regaining self-control and directly correlating with the

Phase-by-Phase Emotional Configuration Model, which posits that rhythmic cues, such as animated transitions, mitigate anxiety during creative or prototyping phases. Completely customise the onboarding experience and render it captivating. Following a brief breathing exercise, users are encouraged to articulate their short-term and long-term aspirations, enabling the app to tailor content addressing stresses like work-related anxiety or sleep disturbances. An approach of 'emotionally granular customisation' demonstrated a 50% increase in user engagement compared to random onboarding methods.

Headspace possesses empirical evidence supporting its efficacy. A study including 1200 individuals experiencing moderate-to-severe stress (Perceived Stress Scale [PSS-10] score >20) is scheduled for 2024. Subgroup analysis revealed a substantial decrease in perceived stress of ≥32% following eight weeks of consistent usage. Engagement data indicates that users who interacted with features such as 'Daily Calm', which displays colour gradients transitioning from red to dark blue, exhibit retention rates more than twice those of those who did not engage. This aligns with experimental findings indicating that enhanced colour transitions facilitated a more rapid emotional recovery from frustration stimuli. This software employs gamification through accomplishment badges and run streak possibilities to continually engage users and stimulate dopamine release. Headspace assists users who do not meet their meditation objectives with sympathetic error management that offers gentle motivational feedback instead of punitive measures, a strategy demonstrated to decrease task desertion among design students.

However, Headspace has several limits in its methodology. However, the cultural universality of the established correlations between colour and emotion is still debatable. Research indicates that certain East Asian individuals are more inclined to perceive white or empty areas as 'calm' rather than blue, underscoring the necessity for adaptive emotional design tactics in educational curriculum. Headspace exemplifies the structural integration of emotional intelligence inside digital goods, serving as a valuable resource for design instructors and students in the development of empathetic, human-centred solutions.

6.4.2 Case Study 2: Nike+ Running App (Gamification and Social Motivation in Fitness Technology)

For every apocalypse, there exists an application, and Nike+ Run Club (NRC) has transformed the fitness application marketplace by converting running from a lonely activity into a live social interaction fostering emotional ties (Rodrigues, et al., 2021). NRC, having surpassed 50 million users, exemplifies the utilisation of gamification and community for sustained engagement, addressing a prevalent challenge in emotionally charged interactions. It also provides insights into churn rates, indicating the frequency with which initial users return daily. The app's success is attributed to its adherence to experimentally recognised frameworks of incentive, habit development and social interaction.

The emotive design of the NRC is predicated on four primary gamified methods. The application starts with progress bars and milestone tracking ('85% to 5K Goal'), using the goal gradient effect, a psychological phenomenon wherein motivation intensifies as users approach their objectives. This is particularly beneficial for maintaining engagement during the 'mid-project slump' phase when individuals tend to get sleepy. Leaderboards and communal run maps are components of the social accountability mechanisms that NRC identifies as its 'visible effort ecosystem', which correlates with a study by Penn State researchers who posited that witnessing such evidence, like community members walking or joggers donning t-shirts that showcase their fitness achievements, motivates others. This fosters healthy competition and peer evaluation, which in both physical fitness and design education is recognised to boost motivation and diminish procrastination. Most NRC users state that they engage in running to prevent social alienation, a phenomenon that might be described as 'competitive empathy'.

Finally, it incorporates dynamic rewards: announcing changeable benefits (for instance, complete a specified number of runs to receive a complimentary audio guide). This aligns with 'The Emotional Surprise Index', which elucidates that randomised positive reinforcement fosters curiosity and inventiveness, hence maintaining player engagement. Groupmuse, a business established by disenchanted conservatory graduates in reaction to their generation's detrimental classical employment opportunities and disdain for the arbiters of high cultural taste. All emotions included within encompass every aspect, submerged inside a singular void of watchfulness. The NRC incorporates biometric emotional mapping directly linked to wearable devices like the Apple Watch, enabling the recommendation of optimal runs depending on real-time stress or tiredness levels. These adaptive responses coincided with emotional-tracking experiments, indicating that physiological recordings might enhance self-regulation and perhaps contribute to the formulation of personalised management solutions. This idea is well endorsed by empirical research investigation revealing that participants in social challenges logged over double the monthly mileage compared to non-participants (Lopes et al., 2024). A 22% increase in happy emotions was seen from publicising personal successes on social media, supporting the concept that 'social validation' acts as an emotional catalyst. NRC's tailored coaching programmes reflected studies indicating that detailed emotional guidelines aid design students in project completion by decreasing dropout rates.

The NRC offers many pedagogical approaches for design education. Commencing with the 'First Run' lesson, users are allocated a virtual coach and directed via animations that illustrate more advanced technology-driven recognition of fatigue, akin to the critiques often seen in most design studios. Short-term challenges like the 'Global 5K Weekend' aim to exploit loss aversion, which has been demonstrated to enhance task prioritisation behaviours in treatments for school children. Furthermore, since designers appreciate tangible results in the form of visuals, progress bars and updated elevation graphs after run maps, these elements may serve as a potent motivational tool for the imposter syndrome that we all experience. Nonetheless, NRC's emphasis on competitiveness may exacerbate sentiments of envy, which manifested in 18% of students during peer review activities. Design must adhere to 'emotional balance guidelines' to ensure the healthiness of incentives and avert a detrimental psychological reaction.

6.4.3 Overview/Significance for Literary Instruction

The Phase-by-Phase Emotional Configuration Model outlined in this book is substantiated by evidence from both Headspace and Nike+ Run Club. Utilising relaxing hues such as blue and green, together with straightforward micro-interactions throughout the app's brainstorming phase, aids in alleviating cognitive strain for users. The progress monitoring and social functionalities of Nike+ maintain motivation throughout the prototype phase. Headspace incorporates non punitive feedback systems that offer critique while facilitating experimentation and iteration, therefore diminishing anxiety levels and enhancing resilience. Informed by these observations, we offer curricular recommendations that incorporate instruments for monitoring emotions and culturally adaptive frameworks, while emphasising the need for assessing emotional alignment in student designs. By examining these applications in real-world contexts, design educators and students may more effectively understand how emotional principles manifest and support practices that foster more compassionate and efficient creative processes.

6.5 SUMMARY

Based on the participating designers' points of view, the design process is significantly influenced by emotion (Ho & Siu, 2012). Junior design students must be taught how emotion might affect decision-making and even final design outcomes. The design approach does not consider all emotions. Common emotions are observably involved in decision-making and cognitive thinking processes during the design process. The main factors influencing these emotions are information processing and decision-making. These two aspects inspire motivation and shape subsequent decisions and behaviours. The commonly observed emotions can be divided into five categories: neutral, external positive, internal positive and internal negative (Oliver, 1993). Throughout the design process, dominant external factors, such as material allocation, ensure that the stimulation by outside sources is greater than that by designers' considerations. Internal influences, such as the personal tastes of designers, place more emphasis on human consideration than environmental stimulation. Neutral emotions are reactions to fairly even exchanges between outside stimuli and internal thoughts. Positive emotions include pleasant or desirable situational reactions, ranging from curiosity and satisfaction to love and pleasure. Designers may experience negative emotions that are uncomfortable and depressing. These emotions can undermine designers' confidence and self-esteem, making them feel negative about themselves and others and generally damaging their sense of well-being.

Depending on the designer's experience, their research and their observations of themselves and their target audience, typical emotions that arise throughout the design process can fluctuate. For instance, curiosity, a component of positive internal feelings, can replace excitement, a component of outward pleasant emotions, once it has been processed. As a result, designers may become curious about what is unknown and the nature of things. Additionally, anxiety, a negative feeling that is a component of neutral emotions, can be processed and replaced with tranquillity, a quiet mental state. When designers find themselves mentally overburdened with worries and to-do lists, they may find it helpful to switch to calmer ideas. Changes in thoughts and emotions' can make design outputs stronger or weaker. Emotions and rational thoughts play important roles in the design process. Balanced decisions and design methods result from balanced approaches to important concerns. Designers' experiences with tools that support analytical thinking and emotional control during the design process have a bearing on their research on design processes. The five categories of emotions are based on a balanced approach that considers reason and emotion. The possibility of including both thoughts and emotions in the design process was studied in this research. The connections between emotion, reasoning and emotional control can contribute to the knowledge on the way designers' emotions arise, change and relate to one another. These findings of the current study can help shape future studies on design and emotion and serve as a starting point for designers to be aware of and observe their emotions.

7 Conclusions

7.1 OVERVIEW

In the 2000s, the efforts of dedicated design researchers who focused primarily on user perspectives stimulated new interest in the correlation between emotional factors and design. However, few studies have examined how junior design students' emotions change during their innovation process or how these emotions affect their creative processes and results. Junior design students may struggle to interpret data that require a high level of human interaction. This implies that the process of creativity involves more than just logical and reasonable considerations; it also involves the method, experience and talent of the designers, as well as their reactions to their immediate surroundings. Designers have emotional reactions to the outside world. As a result, 'intrinsic elements', such as emotions or emotional systems, may influence the aspects that junior design students consider throughout their design process. Theories concerning the interactions between emotions, the decision-making process and the design process have been established by previous research. In this study on the 'design and emotion' element, the intimate connections between designers, design outputs and users were addressed. In addition, the correlations between the emotional shifts junior design students experience, internal variables and external aspects of the design process were examined. By applying the ideas resulting from this study, junior design students will be able to manage the emotional shifts they experience during design processes for improved performance. Potential applications of these ideas include design education. It is necessary to introduce these ideas to junior design students and to ensure that they are able to apply them. Ways to achieve these objectives have yet to be identified.

To help junior design students manage their emotions throughout the design process, this research considered several methods. It is necessary to identify the components of any practical recommendations to use them as viable assistive tools. Therefore, it is important to investigate effective strategies for encouraging younger design students to understand these rules of handling emotional concerns in the design process. It is necessary to consider the challenges of integrating design and emotion principles into design learning. The contributions that can be made by adhering to the rules and incorporating emotional notions into design studies must be put forward. Empirical research was carried out to gather both quantitative and qualitative data for subsequent analysis to accomplish these goals. The suggested concepts were experimented with and used by junior undergraduate participants in creative arts (design) programmes. Two investigations were carried out for this empirical study. The ability of junior design students to control their design processes was improved by inviting contributors to participate in a variety of design methods for studying emotional factors. An emotion-tracking smartphone application kept track

DOI: 10.1201/9781003747345-7

of participants' performance. Individual interviews and an emotion detection system were used to collect their opinions on these suggested concepts. Additionally, data from a focus group were used to assess how well the suggested concepts might work in the collaborative design process. Finally, participant lablogs reflected the factors that may encourage junior design students to impact their design processes using controlled emotions. The purpose of this study was to show junior design students how to incorporate emotion into the design process. The exploration of the relationship between emotional concerns and design emerged throughout the 2000s via the research conducted by design scholars, primarily emphasising the perspectives of users. The design process encompasses not just rational and logical deliberations, but also the designers' skill, expertise and aptitude, along with their responses to the external environment or stimuli that surround them. While these theories provided a sufficient theoretical foundation, there has been little study exploring the fluctuations in emotions experienced by junior design students throughout the design process. Junior design students may find it challenging to absorb knowledge from sources that include a significant degree of human interaction. The significance of incorporating these concepts into the creative processes of junior design students and influencing their manipulation was clearly shown. Nevertheless, there is still a lack of documented solutions to address these problems.

A persistent obstacle has been the students' inability to apply the knowledge acquired in lecture courses on human-centred topics, such as the ageing eye and universal design, to problem-solving in design studio assignments. Design students can exhibit a sense of being overwhelmed when engaged in problem-solving, displaying a lack of ability to identify crucial aspects of ageing and an inability to generate satisfactory solutions, while having completed at least one whole semester of lecture-based instruction on these subjects. In response to the perceived failure of knowledge transfer from lectures to studio problem-solving, as well as the need to educate junior design students about the visual and physiological challenges faced by older adults with eye diseases and vision problems, the researchers aimed to investigate effective methods of education.

We are not alone in our pursuit. The education of young designers has consistently been a cause of discontent among design scholars (Borrego et al., 2010; Cross, 2001; Dorst, 2006; Fitch, 2016; Goldschmidt, 2001; Oxman, 2001; Shadle et al., 2017; Vaughan et al., 2008).

The purpose of this research was to evaluate several techniques for assisting junior design students to maintain emotional control throughout their design process. Practical recommendations are feasible tools to assist them, and the aspects that make up these guidelines must be defined for them to be effective. Accordingly, it was necessary to investigate effective ways to inspire junior design students to master the requirements. To accomplish these goals, empirical research was carried out to gather both quantitative and qualitative data for subsequent analysis. To modify the suggested principles in their design processes, junior undergraduate students in creative arts (design) programmes were asked to participate in the study. Using an emotion-tracking smartphone application, participants were asked to participate in several design processes, and their performance was tracked throughout the process. A comparison of their performance and data related to their emotional shifts was conducted to assess their skills in manipulating the creative process while simultaneously regulating their emotional responses. According to the findings of the empirical investigation, junior design students can incorporate emotion into the design process.

This study investigated approaches to guide junior design students to control their emotions in the design process. Practical guidelines are possible tools that can help them, and the aspects involved in these guidelines must be identified. Effective approaches that can be used to motivate junior design students to learn these guidelines should be explored.

7.2 LIMITATIONS

7.2.1 The Impact of COVID-19 on Conducting Research

Several studies have examined how individuals express strong feelings and how they relate to intense emotions at different times, such as the COVID-19 pandemic (Conti et al., 2020). A general questionnaire was completed anonymously by participating families. In some cases, the children or adolescents suffered from neuropsychiatric disorders. The study investigated different aspects of the daily lives of those young patients and their families during lockdown (Conti et al., 2020). Following the outbreak of the COVID-19 pandemic, new conditions involving online digitally mediated learning were created for educational institutions worldwide. This method may limit the extent of information and customisation achievable from the data. In addition, it is possible that certain families included in the study may have been affected by neuropsychiatric disorders, which might introduce biases or limitations to the results. Moreover, the COVID-19 pandemic necessitated the implementation of social distancing protocols and a shift towards remote learning, leading to novel circumstances for educational establishments worldwide. The practice of social distancing and online classes created new opportunities and challenges that had to be overcome to conduct educational activities and the general teaching process. The latest data suggest that developing and applying electronic communication approaches was unavoidable. This presented opportunities and challenges for engaging in educational endeavours and conducting research. Researchers had to adapt to using digital communication methods and online platforms, potentially impacting the nature and calibre of the gathered data.

7.2.2 Vague Wording of Emotion Descriptions

Words describing emotions in everyday language can be ambiguous. Vague wording can be divided into three main categories: (1) vague, (2) semantically ambiguous and (3) meaning incomplete. Other words used in daily language can be unclear as well. This limitation possibly influenced the veracity of the research findings and the generalisability of the research results. Some inadequate words were used by participants when describing their emotional changes during the design process. This may have been caused by a language problem. The use of words is always accompanied by a specific context. We can understand this context as involving (1) the speaker (message sent out), (2) the hearer (message received) and (3) the occasion of use. Based on these aspects, speech should be clear and precise; however, how clear and precise it is will depend on different contexts. The meaning of a word may be clear in some contexts but not in others. If language is used in relation to a specific context but is not clear in said context, the intended meaning may be obscured. However, even if the participants did not explain themselves clearly, it does not mean that their remarks may be read incorrectly. The research team reviewed related content references to clarify if existing research explained problems related to seemingly ambiguous remarks or asked questions about problems that may arise in this regard. In addition, the research team observed the focus group closely to help the participants describe emotion-related words clearly.

7.3 GENERALISABILITY OF THE RESEARCH FINDINGS

In this study, the entire research process was divided into two Research Studies (1 and 2) that included a design process, a lecture to introduce the concept of design and emotion, a focus group, the use of an emotion-tracking mobile application, an interview and the use of the CAPTIV L7000-Premier application (an emotion-tracking device set) to collect various types of data.

The exploration of the correlation between emotional considerations and design dates back to the early 2000s, during which several design researchers conducted studies primarily focusing on customers. The design process incorporates not just rational and logical elements, but also the techniques, experiences and abilities of designers, as well as their responses to external stimuli. While these theories offer a robust theoretical foundation, limited research has been conducted on the fluctuation of emotions among junior design students during the design process. Junior design students may find it challenging to absorb information obtained from various interactions with people. The need to bring the ideas into the design processes of junior design students and influence the manipulation of these processes was identified. Based on the feedback collected from the focus group, 40 emotions are commonly evoked during the design process. By gaining a deep understanding of these 40 emotions, guidelines can be proposed to help junior design students manage their design processes effectively. However, successful strategies to achieve this objective have not yet been discovered. Accordingly, additional investigation is required.

7.4 FURTHER RESEARCH DIRECTIONS

7.4.1 Emotional Word Kit for the Design Process

As design educators, we can help students create an emotional word kit as a tool to use in the design process. The participants became curious about the differences between emotional words. This improved their literacy and helped them consider healthy emotional responses to events. An emotional word kit may encourage students to be supportive and help each other when they are coping with complex design process tasks or when they get stuck or seem lost. We can encourage students to keep an emotional word kit in their planners. The word kit can occasionally be expanded using an online thesaurus. We can show the thesaurus as an inspirational tool. We can look up some emotional words and discuss alternatives. In addition to using a list of emotional words from online resources, we can put words on cards. Students can group them and turn them into hierarchies, opposites, pairs or a spectrum. Words will change their relationships to one another depending on context. Students typically say something such as, 'Well, it depends on what is happening'. This can serve as positive scaffolding; what we want to encourage is a nuanced approach to context.

Glossary

Design students: Learners who are enrolled in higher education design studies.

Emotions: Emotion is considered a pattern of reactions to both external and internal stimuli. It is also a process that engenders cognitive activities and subjective feelings.

Emotional changes: Changes in emotion (e.g., from negative to positive).

Emotional elements: The effects of various aspects (e.g., sociocultural influences and cognitive processes) that motivate the power of emotion to assign personal relevance and meaning to learning.

Emotional guidelines: Guidelines to help students manage their emotions.

Junior design students: Junior design students are those who have studied design for less than two years.

Appendix

TRANSCRIPT OF FOCUS GROUP DISCUSSIONS

Participant 1: I personally think that the two emotions of curiosity and interest will be most effective and [will help me] achieve my expected learning results.

Because it is only when I am interested in this kind of thing, I will first have the motivation and hope to live up to the results, so I have the power to do it first and keep trying to design my homework and collect all kinds of information. A curiosity, if you are interested, it won't be hard to do it.

Participant 2: I think both are anxious and nervous. Because I don't feel so nervous at the end, so I'm so nervous first (to push myself to learn).

Immediately I felt that there was a design and I felt that I was nervous and anxious and less anxious, because maybe the teacher called me to hand in my homework, and you were shocked to deduct points and stay in a hurry. When it comes out, it's awkward to the point when it comes out.

Participant 3: I think that when I learn something that I am interested in [and] talk about something in class, I am interested in this kind of design, and I am enthusiastic about a certain design. If you are a girl, you will push yourself to learn; that is, you will make yourself want to learn more, and if you have homework at the time, it will help you learn.

Participant 4: I think my excitement will help me learn a lot. Because Xing told me that my mental state would be good, it's not easy for me to look at it, and because my eyes are so difficult to get my own attention, this kind of emotion will make me learn faster.

Participant 5: I think the time when a person settles down, or calms down, is better because I will concentrate. I think a peaceful and calm environment is better because if there are too many people on the side, I don't focus on the side but I'm alone, so I'll be able to understand more ideas.

Participant 6: I think that I can communicate with other people in terms of emotions and maybe be more positive. If I communicate with other people more or talk about situations, it will be easier to learn.

Participant 4: I feel happy and excited to listen to my hyper theory, and I will be more or. I feel that anxiety will be effective for me because when I am under stress, I [am] anxious and want to finish something quickly. I will soon achieve it. Target.

Participant 5: I feel that a [calm] emotion can [create] great learning results because when I am calm, I can [absorb] more knowledge.

Participant 1: I feel that there is a difference between Cherry and me. In my case, personal comfort is the first priority, while Cherry is able to cope with life first. It seems that I have been forced out of some of the ideas, so I prefer to go comfortably, find a point of interest, and then go in-depth and do my homework.

Participant 2: In fact, I am not interested in what kind of things, but it may not necessarily achieve your expected learning results, that is, you usually don't practice, but when you are going to hand in your homework … so nervous In a moment, an IDEA may suddenly be forced out because it is too rushed and shocking, and [you] may immediately return to the hall and give out notes. You can take a sample to make the idea present more completely.

Participant 3: I feel that you must have enthusiasm first to achieve the expected learning results, because you must have goals, first have the motivation to move forward and have motivation, so important, maybe a lot of emotions can lead you to do it. For example, is it okay to be happy?

Or happy? It's okay. But unhappy gold, such as your failure or frustration, can induce a motivation to wait for you to do well. I think …

Participant 4: I agree that euphoria is effective because it can make you more involved in the learning process. I think it's important to be invested.

Participant 5: I think that Calvin's anxiety is to help and effectively accomplish one thing, but if you study, it seems to help, not too big, that is to say, anxiety can help, but it may not be learning.

Researcher: In other words, whether the topic is good or interesting, it can actually help students achieve a learning outcome. Let's take a look at the second article. How do you evaluate the feedback from your teammates during the discussion?

Participant 5: I will vary from person to person, so see the next classmate who gave me my feedback. I will stay with me. I have impressions and judgements [of] my … abilities and personality. Then, according to him, I will go back and estimate how much I will absorb. So, it's not everyone's words, I just take them all as they are. That is, if his talk about construction is constructive and well-founded, I will listen. But if I think it's a free rider, maybe my abilities are not good for me, so I may not listen to it, and I will ignore it.

Participant 1: I personally attach more importance to my teammates because I think, since it is teamwork, I should try my best to cooperate. Everyone has to cooperate with each other, and you will cooperate to achieve what you want to do. I have a project that I want to do.

Researcher: In other words, all three of you listen to the opinions of your classmates, but it is based on the level of the talk, and you will be able to see if you are a credible person or capable, for example, if it is a free rider Oh, it's silly. It's not reasonable.

Participant 2: Before, I had homework to project, don't remember to solve the problem, there is a classmate who will go back to school if he rarely goes back to school. So, my other classmate and I will stand by and kick ya. Another classmate is so fair. Well, I think, my classmates are so smart, so Creative can stay low.

So, I look at it again and again. Actually, it's a lot of energy, so I'm just fine, so I'll stay back.

Researcher: Does the result achieve the goal you want to achieve?

That is, it is okay, but the workload needs to be done less.

Participant 3: I think … proactive investment is very important for all learning.

Participant 4: Because I'm in a lot of emotions, it's always about three minutes when I'm feeling about things. I've learned the concept quickly, but if I follow along, I'll drink quickly. It will be so clear that mistakes made will be corrected quickly.

Sometimes I think about the concept so quickly, but if I follow it, I can enter it quickly. I don't remember. So if I have time to settle down, my thoughts will be [clearer] and mistakes [can] be corrected quickly.

Researcher: The other one has nothing to add because of a classmate?

Participant 2: I think the psychological quality is so important, and the mentality is right when I share my homework, because if it is really mean, it's like talking with others, it's actually not a favourite of people to think that they're wrong, maybe they'll think of me. It turns out that he understands the other party's first and then talks about it. Listening to the side, I can suggest that it will be more comfortable to listen to you, and it will be much better to bury it into a rundown.

Participant 4: If you understand the other party, you can talk about it before you listen to it. There are some places you can suggest. It will be more comfortable if you listen to it, and it will be much better if you bury it in a rundown.

Participant 5: I think that sometimes the topics are very subjective, maybe the same artwork may have different answers to the human body, I will respect a teammate's approach, or I will consider or try it. Talking about the Dharma system.

Researcher: How about the other student?

Participant 3: I think it depends on how a friend is more than if it is really an opinion, then I will listen to it. If there is something to think about making trouble or uploading you, I think that love will be the wrong way.

Participant 1: I will be rational to observe first, rationally discuss.

Researcher: Let's move on to the next one. The third one asks you, 'What kind of feedback do you think is more effective for your study?'

Participant 1: I think the first feedback is not too long, it's about three sentences, because if it's not me, it will be hard to remember, if it's not, maybe I will write down the paper, it's not me Will be so fast, don't remember. And if it is perfect (to a concept), I don't think I would mind too much difference (not to talk about his attitude), that is, if my work is so bad, it can be as good as it is. I will think that the completed comments are true words (helping to learn effectively).

Participant 2: I'm like a classmate who can tell the truth, but it's not too MEAN, you can go around and talk about it. It sounds like encouragement, it's helpful first.

I'm the same head, I'm not much different from me, I'm always accepting the kind, even if I'm really good, I'm always inclined to beautify well or talk good, it's better than … it's better than [when] you are irritable, or the other person is in a good mood. It's bad, so I think it's just a matter of venting (emotions), and I receive information that is not valuable, even if there are some points worth referring to or the information is because of their emotions or their useful information. If you're emotional, follow me and I'll accept it until I'm shutting it down. It's a situation where you don't want to be too sensible, and I won't listen.

Researcher: So, do you think there will be feedback from the group? Do you think it can help you? Is it so gentle? Isn't it sharp?

Participant 1: I think it's not emotional if there is a comment.

Participant 2: I think my classroom and classmates stay and comment are good friends.

Participant 3: I will see if it is useful if it is typed in. If it is not good, then it can improve my words, so I think they will listen to it.

Participant 4: If it's my opinion that I don't agree with, I'll look at it. I'll use the form to express it. If it doesn't work at all, I won't see you again. I'll look at it after it's adopted. Then it will be improved to the point where I will listen to it.

Participant 5: I haven't tried a good experience, it's what my classmates say.

Participant 6: I usually work as an intermediary, and I don't want to talk to my classmates. I think the most difficult thing is that both of them are trying to stand on their own. If they feel awkward, because there are some people, they will feel that their position will be good, and they will not listen to others. If it happens that two people are in the same group, Zhonghui will not play the role of teammates. I think that sometimes the deadline for oneself is not necessarily a good thing.

Participant 2: But when there are some unreasonable troubles, I don't even have a reason to go because it's because they may be in the wrong direction, then it's up to the location and it must be adjusted appropriately.

Participant 3: Because my personality is relatively good, it will only be part of the agreement, and it will be a tactful way to talk to him. In fact, it may be better than just being rational.

Participant 4: I think I will use it if I think it's okay. If I think it's not okay, I won't use it, it's not complicated and I would like to keep it simple.

Participant 3: When there is an array, if he sends an idea, it will be a new suggestion. You will ask for it. It may not be the main reason. You will always look back in which direction. There will be a chance to improve.

Participant 1: When I receive the comment for the first time, I will reflect on my opinions. Many times, the opinions may not necessarily be understood [right away]. The first reaction is to reflect immediately. I will consider it. The feedback is true, and I will immediately consider the matter, but sometimes I may not have this problem, but I will believe it. First, sometimes I will immediately look at the other person's perspective to see myself and I don't remember my position.

Participant 3: [Many] times, I just asked my classmates to give me some feedback. The feedback is so constructive. Most of them are just what you want. Okay, I think it can be more MEAN.

Researcher: If you have a bleating emotion, you will be happy when you receive the feedback?

Participant 2: To view the feedback content.

Participant 3: If the teacher said [it], I might question myself. There are so few classmates who are not too doubtful.

Participant 1: If … homework is a topic that I am interested in, then I will be proactive at the beginning, but if I understand it, I will go to the very end and do it first. Talk about a deadline.

Participant 2: If I am in a normal mood, then my design process will progress normally.
But if I'm not happy, I don't even think about it until the design will end. If I'm hungry, I won't even go to the design.

Researcher: Do you know if there is a way to maintain your mood until your mood is normal? Is there a way or a secret technique?

Participant 2: UM … Maybe watch the pretty boy or listen to K-pop.
If I am unhappy or uncomfortable, it will be very difficult to concentrate and bury and want to give up, or if I'm late, I'll do it first. And burying the deadline is really effective, so it makes me move forward.

Participant 2: I think it's unhappy, it's very ugly to talk about it.

If I'm in a good mood, I'll try to do it. I think I can do it, but if I'm tired of that kind of thing, I'll do it.

Participant 3: I think that the positive is many times better than the negative because every time you make a work, there may actually be feedback. You make yourself improve, but if you have more positive feedback, it will give you a big boost. In fact, the origin of everything you do is your own driving force. So, I think positive feedback is to promote the most fundamental direction of your doing.

Participant 4: I think that positive feedback is so important to an action or a motivation to continue to make work. [But] negative [feedback] is as important as a comparative comment. Will it be true? Will it be true, but maybe we will be so negative? Look at it.

Participant 5: I personally think that positive feedback is better. Negative feedback is … not good, because negative feedback is all because some people say that it may be a kind of radical method, which can be used [as a] radical method to push individuals to do better.

Participant 6: I think the positive and negative should be [equal] because if it is too positive for [too long], you will not think about your own direction. In fact, it may be wrong, but if it is negative, it may not have a driving force to make me. I think the positive and negative effects [should be balanced].

Participant 1: I think that if you have a small amount of negative, point out the problem and then use the positive one to talk back to you. You know how to calm down the matter, and the solution will be effective because many times you may not know that you are sitting on the sidelines. A negative one can make you know that you are sitting on the sidelines.

Participant 2: I think positive is better than negative. If the premise is that the negative is more effective … I think it needs to be accompanied by constructive suggestions. The previous question is that the suggestions are really helpful. But personally, I want to make a more positive return to make me feel that there is more to learn.

Participant 4: There will be thoughtful emotions because I feel that everyone is too young and should look at the law, and they should be too different. If I receive other people's opinions that are different, I will continue [search for] the truth. Why [am I not] the same?

Participant 5: I think positivity will certainly make me happy and motivated to [create] a piece of work. Negative I think both will be unhappy, but they will really reflect on whether there is a way to do it well or how to do it well. It is recommended that you [create] another work and the [result] can be improved.

Participant 6: I have always thought that feedback is positive. It is better because you may be better. You did it [and] it was hard work, but it was all negative. It would be so unhappy.

Participant 1: If you receive a negative comment, you will feel unhappy, but you will see it again.

Participant 3: I think anxiety is not good for the work, because negative emotions will make you feel about your own work. The way you look at it is not the same, as I think positive or negative emotional changes will affect the final success of your work. Let's take a look at your work.

Participant 4: I feel confident that the process is so important, as if [Photoshop skills are] not enough, I will just give up directly. I will not continue to use this method. I don't know if you have tried to use this one, but it is. Actually, I really learned Photoshop a long time, so I have so much motivation to make a picture or learn the software.

Participant 6: Sometimes when I am sad, the element I use is not wild; it's only grey, it's a single hue, it's not a lot of conversation, and colour isn't used a lot. The hue may be only black, white or grey.

Researcher: After you reflect and digest, how long does it take to be understood (e.g., the impact of emotion on design).

Participant 1: There may be something, sometimes it's really awkward, but it's possible, sometimes you'll understand with a little bit, but it may be the problem, maybe I really have to do it many times, I live there many times, maybe it's two or three months, and the teacher may talk about it. I'm going to say the same thing, but I want to get there first after two or three weeks.

Participant 2: Sometimes it's necessary to be patient and … realise that when there is a [problem], people will be so much worse, and then I will watch the training field. If the coefficient is learned, it may take a day or two to understand it, or if there is a problem, I think it is possible. I've never understood in my life, I think, but if it's a design, it's freer, and maybe you tell me about it when I'm on the road, when I'm free every day. How much time is it, I will understand it.

Participant 3: I think that if we learn by ourselves, we can spend a lot of time reflecting, because it is not [a question] that has an absolute answer … In fact, it is not the same as the one that has the correct answer … that is. Then you will have theories to generate other ideas, so you will need to reflect on it.

Participant 4: I usually spend one night reflecting on it, because it's clearest before I feel it. I usually ask for time to reflect on my reaction.

Participant 5: I usually use shower time, because shower time is like taking me to my adult life.

Participant 6: The time for my reflection may be the time when I [am on a] bus. When I [am on the] bus, I might look at the scenery before I … reflect on what I learned.

Participant 3: I think the time for reflection may not be so stable, so that few people will sit in the bottleneck position and do not know how to do it tightly. In fact, I will spend some time reflecting on the possibility of tightness. Just take a few minutes or think about it [for] 10 minutes, just think about it

Researcher: Oh, yes! We missed the last question.

Participant 3: Actually, I think that emotions may have an impact, but I find it difficult to learn or [find] time to learn. Although I think … that is, you recommend [something to] me, and every app is good. I will do it in class, but in life, a person may not have good self-discipline.

Participant 1: I think emotional management and reflection are very suitable for me because I understand that I am emotion, so it is easy to be affected by others, so emotions will be difficult for me to grasp the fluctuations and have a detailed teacher. Parents will always persuade me. For example, if I am not a good person, I don't need to converge when I am in a battle. I have to say it and I will be straight to express it. If you tell me, I will have a reflection first. It's because people talk about it, and I will remember it, and then I'll go back to the people's talk about eating and eating, and burying emotions and mood swings will affect my work efficiency.

Participant 2: I think (not enough emotion management in the classroom and using apps) should be suitable, that is, I have said that if my emotion is bad, I will do it not enough to provide my work efficiency should I have to learn about emotional management to make me more efficient. Do more and speak out. Reflection is important because if there is no reflection, there will be no progress, and people have to improve.

Participant 1: I want to say that reflection will make me all, but I can talk about it.

Participant 4:	I think emotions are actually difficult to manage because at the moment you feel personally frustrated, it is difficult for you to tune it directly. It is a frustrating emotion that may be tuned before you can do it well because it actually records you when you are. It's a moment of feeling, so it's profound to make a singular work.
Participant 5:	I think emotional management and reflection are both important because if you are a negative one, you may be so negative, and if you use reflective emotions, you will be so rational. So, I think Fan Xiao has to do it before. Manage your emotions well, so you will reflect on first come first.
Participant 6:	I think emotions and reflections are very important. Maybe it's a good hyper for one night, but I don't know what to do. I think it's good to stay with you until the second night or after a while and then come back and go to think about you. What is it? In fact, I think that sometimes I feel so happy. Sometimes I don't necessarily know what I'm doing, so I need time to reflect on it to see whether it's really good or bad.

References

Achmad, A. D., Qotadah, H. A., & El Qubro, H. S. (2023). Analysis of the correlation between gratitude and emotional intelligence among university students. *Journal of Education Method and Learning Strategy, 1*(03), 166–173.

Alaniz, T., & Biazzo, S. (2019). Emotional design: The development of a process to envision emotion-centric new product ideas. *Procedia Computer Science, 158*, 474–484.

Al-Jedaia, Y., & Mehrez, A. (2020). The effect of performance appraisal on job performance in governmental sector: The mediating role of motivation. *Management Science Letters, 10*(9), 2077–2088.

Altay, B., Ballice, G., Bengisu, E., Alkan-Korkmaz, S., & Paykoç, E. (2016). Embracing student experience in inclusive design education through learner-centred instruction. *International Journal of Inclusive Design, 20*(11), 1123–1141.

Ambrose, G., Harris, P., & Ball, N. (2019). *The fundamentals of graphic design*. Bloomsbury Publishing.

Anderson, L., & Krathwohl, D. (Eds.). (2001). *A taxonomy for learning, teaching and assessing: A revision of Bloom's taxonomy of educational objectives*. Longman.

Arias, A. M., & Davis, E. A. (2017). Supporting children to construct evidence-based claims in science: Individual learning trajectories in a practice-based program. *Teaching and Teacher Education, 66*, 204–218.

Assari, M. A., & Rahmati, M. (2011). Driver drowsiness detection using face expression recognition. In *2011 IEEE international conference on signal and image processing applications* (ICSIPA, pp. 337–341). IEEE. https://doi.org/10.1109/ICSIPA.2011.6144162

Atkinson, R. C., & Shiffrin, R. M. (1968). Human memory: A proposed system and its control processes. In K. W. Spence, & J. T. Spence (Eds.), *Psychology of learning and motivation* (Vol. 2, pp. 89–195). Academic Press.

Awang, H., & Ramly, I. (2008). Creative thinking skill approach through problem-based learning: Pedagogy and practice in the engineering classroom. *International Journal of Human and Social Sciences, 3*(1), 18–23.

Azadi, Y., Yazdanpanah, M., & Mahmoudi, H. (2019). Understanding smallholder farmers' adaptation behaviors through climate change beliefs, risk perception, trust, and psychological distance: Evidence from wheat growers in Iran. *Journal of Environmental Management, 250*, 109456.

Badran, O., & Al-Haddad, S. (2018). The impact of software user experience on customer satisfaction. *Journal of Management Information and Decision Sciences, 21*(1), 1–20.

Bahlgerdi, M., Miri, M., Sharifzadeh, G., Norozi, E., & Hosseini, T. (2021). The effect of educational intervention based on theory of planned behavior and self-regulatory strategies on the social vitality of women employee. *Journal of Education and Health Promotion, 10*(2021), 1–7.

Bailey, E., Alvarez-Jimenez, M., Robinson, J., D'Alfonso, S., Nedeljkovic, M., Davey, C. G., Bendall, S., Gilbertson, T., Phillips, J., Bloom, L., et al. (2020). An enhanced social networking intervention for young people with active suicidal ideation: Safety, feasibility and acceptability outcomes. *International Journal of Environmental Research and Public Health, 17*(7), 2435.

Bares, C. T., Morillo-Rivero, L. E., & Papini, M. R. (2021). When loss hurts: Psychobiological basis of frustration. *Revista Interamericana de Psicología/Interamerican Journal of Psychology, 55*(2), e1443–e1443.

Bastian, B. (2017). A social dimension to enjoyment of negative emotion in art reception. *Behavioral and Brain Sciences, 40*, e352.

Baumgartner, C., & Pirker, S. (2019). Video-EEG. *Handbook of clinical neurology, 160* (2019), 171–183.

Benedict, S., & Hilton, K. (2006). *The emotional spectrum analysis 16 EEG system: Practical and conceptual considerations for objectively investigating experienced emotion in design research.* Proceedings of the 5th. International Conference on Design and Emotion.

Bennett, L. (2014). Putting in more: emotional work in adopting online tools in teaching and learning practices. *Teaching in Higher Education, 19*(8), 919–930.

Berman, N. (2020). A critical examination of informal learning spaces. *Higher Education Research & Development, 39*(1), 127–140.

Best, K. (2006). *Design management: Managing design strategy, process and implementation* AVA Publishing.

Bilevicius, E., Single, A., Bristow, L. A., Foot, M., Ellery, M., Keough, M. T., & Johnson, E. A. (2018). Shame mediates the relationship between depression and addictive behaviours. *Addictive Behaviors, 82* (2018), 94–100.

Bixler, J. L. (2016). *Forgiveness: A learned cognitive reappraisal of anger.* The University of the Rockies.

Bless, H., Fiedler, K., & Forgas, J. P. (2006). Mood and the regulation of information processing and behavior. In K. W. Spence, & J. T. Spence (Ed.),*Affect in Social Thinking and Behavior*, Psychology Press.

Bloom, B. S., & Krathwohl, D. R. (1956). Taxonomy of educational objectives: The classification of educational goals by a committee of college and university examiners. *Handbook I: Cognitive domain.* Longmans, Green.

Boess, S. U., & de Jong, A. M. (2007). Communicating user research into new product development processes. In Poggenpohl (Ed.)S. , *IASDR07 proceedings: Emerging trends in design research* (pp. 1–8). The Hong Kong Polytechnic University.

Boess, S., de Jong, A., Kanis, H., & Rooden, M. (2009). *Investigative designing: Usage-oriented research in and through designing.* Sheffield Hallam University.

Bongiovanni, I., & Louis, C. P. (2021). Theory and practice of Design Thinking: Perspectives of designers and business consultants. *International Journal of Design Creativity and Innovation, 9*(3), 174–191.

Bordone, M.P., Salman, M.M., Titus, H.E., Amini, E., Andersen, J.V., Chakraborti, B., Diuba, A.V., Dubouskaya, T.G., Ehrke, E., Espindola de Freitas, A., Braga de Freitas, G., Gonçalves, R.A., Gupta, D., Gupta, R., Ha, S.R., Hemming, I.A., Jaggar, M., Jakobsen, E., Kumari, P(2019). The energetic brain – A review from students to students. *Journal of Neurochemistry, 151*(2), 139–165.

Borrego, M., Froyd, J., & Hall, T. S. (2010). Diffusion of engineering education innovations: A survey of awareness and adoption rates in the U.S. engineering departments. *Journal of Engineering Education, 99*(3), 185–207.

Brew, A. (2012). Teaching and research: New relationships and their implications for inquiry-based teaching and learning in higher education. *Higher Education Research & Development, 31*(1), 101–114.

Brewer, K. L., & Devnew, L. E. (2022). Developing responsible, self-aware management: An authentic leadership development program case study. *The International Journal of Management Education, 20*(3), 100697.

Briñol, P., Petty, R. E., Stavraki, M., Lamprinakos, G., Wagner, B., & Díaz, D. (2018). Affective and cognitive validation of thoughts: An appraisal perspective on anger, disgust, surprise, and awe. *Journal of Personality and Social Psychology, 114*(5), 693–718. https://doi.org/10.1037/pspa0000118

Brissaud, D., Garro, O., Poveda, O. (2003). Design process rationale capture and support by abstraction of criteria. *Research in Engineering Design*, *14*, 162–172.

Brocato, K. (2009). Studio based learning: Proposing, critiquing, iterating our way to person-centeredness for better classroom management. *Theory into Practice*, *48*(2009), 138–146.

Brom, C., Starkova, T., & D'Mello, S. K. (2018). How effective is emotional design? A meta-analysis on facial anthropomorphisms and pleasant colors during multimedia learning. *Educational Research Review*, *25*(2018), 100–119.

Bucci, D. J., Phillips, R. G., & Burwell, R. D. (2021). Contributions of postrhinal and perirhinal cortex to contextual information processing. *Behavioral Neuroscience*, *135*(3), 313–325. https://doi.org/10.1037/bne0000436

Buheji, M. (2019). *Designing a curious life*. AuthorHouse.

Burrows, C. R., & Sahinkaya, M. N. (1985). Assessment of techniques for estimating oil-film bearing coefficients. *Tribology International*, *18*(5), 267–274.

Caldeira, C., Chen, Y., Chan, L., Pham, V., Chen, Y., & Zheng, K. (2017). Mobile apps for mood tracking: An analysis of features and user reviews. In *AMIA annual symposium proceedings* (pp. 495–504). American Medical Informatics Association.

Capps, D. K., & Crawford, B. A. (2013). Inquiry-based professional development: What does it take to support teachers in learning about inquiry and nature of science? *International Journal of Science Education*, *35*(12), 1947–1978.

Carver, C. (2006). Approach, avoidance, and the self-regulation of affect and action. *Motivation and Emotion 30*(2), 105–110. https://doi.org/10.1007/s11031-006-9044-7

Casals-Gutierrez, S., & Abbey, H. (2020). Interoception, mindfulness and touch: A meta-review of functional MRI studies. *International Journal of Osteopathic Medicine, 35* (2020), 22–33.

Chatterjee, A., Gupta, U., Chinnakotla, M. K., Srikanth, R., Galley, M., & Agrawal, P. (2019). Understanding emotions in text using deep learning and big data. *Computers in Human Behavior*, *93*, 309–317.

Chen, A., & Chen, R. (2004). An adaptive design process generated by the integration of systematic design process and design patent protection mechanism. *International Journal of General Systems*, *33*(6), 635–653.

Chen, C. H., Lee, I. J., & Lin, L. Y. (2015). Augmented reality-based self-facial modeling to promote the emotional expression and social skills of adolescents with autism spectrum disorders. *Research in Developmental Disabilities, 36* (2015), 396–403.

Chen, H. T. (2014). *Practical program evaluation: Theory-driven evaluation and the integrated evaluation perspective*. Sage Publications.

Chen, S., Benedicktus, R., Kim, Y., & Shih, E. (2018). Teaching design thinking in marketing: Linking product design and marketing strategy in a product development class. *Journal of Marketing Education*, *40*(3), 176–187. https://doi.org/10.1177/0273475317753678

Christner, N., Pletti, C., & Paulus, M. (2020). Emotion understanding and the moral self-concept as motivators of prosocial behavior in middle childhood. *Cognitive Development, 55* (2020), 100893. https://doi.org/10.1016/j.cogdev.2020.100893

Ciompi, L., & Panksepp, J. (2005). Energetic effects of emotions on cognitions: Complementary psychobiological and psychosocial findings. In R. D. Ellis & N. Newton (Eds.), *Consciousness & emotion: Agency, conscious choice, and selective perception* (pp. 23–55). John Benjamins Publishing Company. https://doi.org/10.1075/ceb.1.04cio

Clinton, G., & Rieber, L. P. (2010). The studio experience at the University of Georgia: An example of constructionist learning for adults. *Educational Technology Research and Development*, *58*(6), 755–780.

Conti, E., Sgandurra, G., De Nicola, G., Biagioni, T., Boldrini, S., Bonaventura, E., Buchignani, B., Della Vecchia, S., Falcone, F., Fedi, C., Gazzillo, M., Marinella, G., Mazzullo, C., Micomonaco, J., Pantalone, G., Salvati, A., Sesso, G., Simonelli, V., Tolomei, G., … Battini, R. (2020). Behavioural and emotional changes during COVID-19 lockdown in an Italian paediatric population with neurologic and psychiatric disorders. *Brain Sciences, 10*(12), 918. https://doi.org/10.3390/brainsci10120918

Council for Interior Design Accreditation Professional Standards. (2018). *Professional standards.* Retrieved May 27, 2019, from https://accredit-id.org/professional-standards/

Cooper, J. O., Heron, T. E., & Heward, W. L. (2013). *Applied behavior analysis* (2nd ed.). Pearson.

Craik, F. I. M., & Lockhart, R. S. (1972). Levels of processing: A framework for memory research. *Journal of Verbal Learning and Verbal Behavior, 11*(6), 671–684. https://doi.org/10.1016/S0022-5371(72)80001-X.

Crilly, N. Moultrie, J., & Clarkson, J. (2004). Understanding product appearance: Cognitive and emotional response to product visual form. *Paper presented at Design and Emotion Conference 2004,* Ankara, Turkey.

Cross, M., & Sivaloganathan, S. (2004). A methodology for developing company-specific design process models. *Proceedings of the Institution of Mechanical Engineers, Part B: Journal of Engineering Manufacture, 219*(3), 265–282. https://doi.org/10.1243/095440505X28972

Cross, N. (2001). Design cognition: Results from protocol and other empirical studies of design activity. In C. M. Eastman, NewstatterW. , & W. M. McCracken (Eds.), *Design knowing and learning: Cognition in design education* (pp. 79–103). Elsevier.

D'Mello, S., Lehman, B., Pekrun, R., & Graesser, A. (2014). Confusion can be beneficial for learning. *Learning and Instruction, 29*(2014), 153–170. https://doi.org/10.1016/j.learninstruc.2012.05.003

Dael, N., Mortillaro, M., & Scherer, K. R. (2012). Emotion expression in body action and posture. *Emotion, 12*(5), 1085–1101. https://doi.org/10.1037/a0025737

Darwin, C., & Prodger, P. (1998). *The expression of the emotions in man and animals.* Oxford University Press.

Dawkins, K., & Furnham, A. (1989). The colour naming of emotional words. *British Journal of Psychology, 80*(3), 383–389.

Dawood, F., & Kashif, M. (2021). My love, I hate you: Rethinking brand jealousy to investigate emotional transition among rural customers. *Strategic Change, 30*(1), 45–52.

De Garrido, L. (2021). Conceptual design of a creative artificial intelligence system based on the neurocognitive bases of human creativity in the brain. *Creativity Research Journal,* 34 (3), 273–294. https://doi.org/10.1080/10400419.2021.2005309

Demir, E., Desmet, P. M., & Hekkert, P. (2009). Appraisal patterns of emotions in human-product interaction. *International Journal of Design, 3*(2), 41–51.

Denton, H. G., McDonagh, D., Baker, S., & Wormald, P. (2004). Introducing the student designer to the role of emotion in design. In McDonagh, D., Hekkert, P., van Erp, J., & Gyi, D. (Eds.). *Design and emotion: The experience of everyday things* (pp. 415–420). Taylor & Francis.

Desmet, P. M. (2015). Design for mood: Twenty activity-based opportunities to design for mood regulation. *International Journal of Design, 9*(2), 1–19.

Desmet. (2018). Measuring emotion: Development and application of an instrument to measure emotional responses to products. In Blythe, M. A., Overbeeke, K., Monk, A. F., Wright, P. C. (Eds.), *Funology 2* (pp. 391–404). Springer International Publishing.

Despeisse, M. (2018). Games and simulations in industrial engineering education: A review of the cognitive and affective learning outcomes. In *2018 Winter simulation conference* (WSC, pp. 4046–4057). IEEE.

Dikmen-Yildiz, P., Ayers, S., & Phillips, L. (2018). Longitudinal trajectories of post-traumatic stress disorder (PTSD) after birth and associated risk factors. *Journal of Affective Disorders*, *229*(2018), 377–385.

Dorst, K. (2006). Design problems and design paradoxes. *Design Issues*, *22*(3), 4–17. https://doi.org/10.1162/desi.2006.22.3.4

Dunne, A. (1999). *Hertzian tales: Electronic products, aesthetics experience and critical design*. RCA CRD Research Publications.

Eastman, C. M. (2001). New directions in design cognition: Studies of representation and recall. In C. M. Eastman, W. C. Newstetter, & McCrackenC. (Eds.), *Knowing and learning to design* (pp. 147–198). Elsevier Science.

Egger, M., Ley, M., & Hanke, S. (2019). Emotion recognition from physiological signal analysis: A review. *Electronic Notes in Theoretical Computer Science, 343*(2019), 35–55. https://doi.org/10.1016/j.entcs.2019.04.009

Ekman, P. (1992). Are there basic emotions? *Psychological Review*, *99*(3), 550–553. https://doi.org/10.1037/0033-295X.99.3.550

Ekman, P. (1994). Strong evidence for universals in facial expressions: A reply to Russell's mistaken critique. *Psychological Bulletin*, *115*(2), 268–287. https://doi.org/10.1037/0033-2909.115.2.268

Ekman, P. (2003). Darwin, deception, and facial expression. *Annals of the New York Academy of Sciences*, *1000*(1), 205–221.

Ekman, P., & Friesen, W. V. (1971). Constants across cultures in the face and emotion. *Journal of Personality and Social Psychology*, *17*(2), 124. https://doi.org/10.1037/h0030377

Endres, T., Weyreter, S., Renkl, A., & Eitel, A. (2020). When and why does emotional design foster learning? Evidence for situational interest as a mediator of increased persistence. *Journal of Computer Assisted Learning*, *36*(4), 514–525.

Ertmer, P. A., & Russell, J. D. (1995). Using case studies to enhance instructional design education. *Educational Technology*, *35*(4), 23–31.

Esser, P. (2018). *How to improve your designs by developing empathy for your target group*. Interaction Design Foundation. Retrieved January 2, 2019, from https://www.interaction-design.org/literature/article/empathy-how-to-improve-your-designs-by-developing-empathy-for-your-target-group

Feldon, D. (2007). Cognitive load and classroom teaching: The double-edged sword of automaticity. *Educational Psychologist*, *42*(3), 123–137. https://doi.org/10.1080/00461520701416173

Fitch, N., & Tarbutt, M. (2016). Principles and design of a Zeeman–Sisyphus decelerator for molecular beams. *ChemPhysChem*, *17*(22), 3609.

Fitch, S. (2016). *Critiques, credits and credibility: Assessment practices in higher education studio art courses* [Doctoral dissertation, Concordia University]. https://spectrum.library.concordia.ca/981013/1/Fitch_PhD_S2016.pdf

Formosa, M. (2019). Active ageing through lifelong learning: The university of the third age. *The University of the Third Age and Active Ageing*, *23*(2019), 3–18.

Formosa, J., Frommel, J., Mandryk, R. L., Tobin, S. J., Türkay, S., & Johnson, D. (2025). Emotion regulation, need satisfaction, passion and problematic video game play during difficult times. *Behaviour & Information Technology*, *44*(13), 3271–3296.

Freud, S. (2012). *The basic writings of Sigmund Freud*. Modern library.

Fried, L., & Chapman, E. (2012). An investigation into the capacity of student motivation and emotion regulation strategies to predict engagement and resilience in the middle school classroom. *The Australian Educational Researcher*, *39*(3), 295–311.

Friedman, B. H. (2010). Feelings and the body: The Jamesian perspective on autonomic specificity of emotion. *Biological Psychology*, *84*(3), 383–393.

Gaver, B., Dunne, T., & Pacenti, E. (1999). Cultural probes. *Interactions*, *6*(1), 21–29.

Gerald, G. R. (2010). *Effects of emotional intelligence on teacher retention*. Southeastern Louisiana University.

Giotakos, O. (2022). The intentional system: Tracing the conative feature of aboutness and directedness in human brain. *Dialogues in Clinical Neuroscience & Mental Health, 5*(4), 143–148. https://doi.org/10.26386/obrela.v5i4.242

Goldschmidt, G. (2001). Visual analogy: A strategy for design reasoning and learning. In C. M. Eastman, W. M. McCracken, & W. C. Newstetter (Eds.), *Design knowing and learning: Cognition in design education* (pp. 199–219). Elsevier.

Greenberg, S. (2009). Embedding a design studio course in a conventional computer science program. In *IFIP International Federation for Information Processing, Volume 289, Creativity and HCI: From experience to design in education* (pp. 23–41). Springer.

Guerra, M. A., & Tripp, S. (2018). Theoretically comparing design thinking to design methods for large-scale infrastructure systems. In *DS 89: Proceedings of the fifth international conference on design creativity* (ICDC 2018; pp. 168–175). University of Bath.

Han, K. H., & Park, J. W. (2009). Process-centered knowledge model and enterprise ontology for the development of knowledge management system. *Expert Systems with Applications, 36*(4), 7441–7447.

Harley, J. M., Pekrun, R., Taxer, J. L., & Gross, J. J. (2019). Emotion regulation in achievement situations: An integrated model. *Educational Psychologist, 54*(2), 106–126.

Harmon-Jones, E., Harmon-Jones, C., & Summerell, E. (2017). On the importance of both dimensional and discrete models of emotion. *Behavioral Sciences (Basel, Switzerland), 7*(4), 66. https://doi.org/10.3390/bs7040066

Harvey, S. T., Bimler, D., Evans, I. M., Kirkland, J., & Pechtel, P. (2012). Mapping the classroom emotional environment. *Teaching and Teacher Education, 28*(4), 628–640.

Heresco-Levy, U., Greenberg, D., & Dasberg, H. (1990). Family expressed emotion: concepts, dilemmas and the Israeli perspective. *The Israel Journal of Psychiatry and Related Sciences, 27*(4), 205–215.

Herrmann, J. W., & Schmidt, L. C. (2002, August). *Viewing product development as a decision production system*. Proceedings of Design Engineering Technical Conferences and Computers and Information in Engineering Conference, Montreal, Canada.

Herrmann, M. R., Brumby, D. P., Oreszczyn, T., & Gilbert, X. M. (2018). Does data visualization affect users' understanding of electricity consumption? *Building Research & Information, 46*(3), 238–250.

Heskett, J. (2017). *Design and the creation of value*. Bloomsbury Publishing.

Ho, A. G., & Chau, P. W. (2016). Understanding the capacity of emotion in decision-making for designers in design process management. *The International Journal of Design Management and Professional Practice, 10*(4), 13.

Ho, A. G., & Siu, K. W. M. (2012). Emotion design, emotional design, emotionalised design: A review on their relationships from a new perspective. *The Design Journal, 15*(1), 9–32. https://doi.org/10.2752/175630612X13192035508462

Hoemann, K., Fan, M., Engen, H., Chou, C. A., Quigley, K., Gendron, M., & Barrett, L. F. (2020). *A network analytic approach to measuring emotional granularity*. Psychology Archive.

Holmes, K. J., Alcat, C., & Lourenco, S. F. (2019). Is emotional magnitude spatialized? A further investigation. *Cognitive Science, 43*(4), e12727. https://doi.org/10.1111/cogs.12727

Horvath, G. A., Tarailo-Graovac, M., Bartel, T., Race, S., Van Allen, M. I., Blydt-Hansen, I., Ross, C. J., Wasserman, W. W., Connolly, M. B., & van Karnebeek, C. D. (2018). Improvement of self-injury with dopamine and serotonin replacement therapy in a patient with a hemizygous PAK3 mutation: A new therapeutic strategy for neuropsychiatric features of an intellectual disability syndrome. *Journal of Child Neurology, 33*(1), 106–113. https://doi.org/10.4135/9780857021021

Howard, M. C. (2019). Applying the approach/avoidance framework to understand the relationships between social courage, workplace outcomes, and well-being outcomes. *The Journal of Positive Psychology, 14*(6), 734–748.

Huanhuan, R., & Chi, M. (2016). Analysis of teachers' role as emotional guide under flipped classroom teaching context. In KotzéP. , W. Wong, J. Jorge, A. Dix, & P. A. Silva (Eds.), *7th international conference on education, management, information and computer science* (ICEMC 2017, pp. 74, 645–650). Atlantis Press.

Immordino-Yang, M. H. (2011). Me, my "self" and you: Neuropsychological relations between social emotion, self-awareness, and morality. *Emotion Review, 3*(3), 313–315.

Jablonsky, R. (2022). Meditation apps and the promise of attention by design. *Science, Technology, & Human Values, 47*(2), 314–336.

James, M. (2017). Advancing design thinking towards a better understanding of self and others. *FormAkademisk – Forskningstidsskrift for design og designdidaktikk, 10*(2). https://doi.org/10.7577/formakademisk.1649.

Jensen, R. (1999). *The dream society: The coming shift from information to imagination.* McGraw-Hill.

Jiang, M., & Koo, K. (2020). Emotional presence in building an online learning community among non-traditional graduate students. *Online Learning, 24*(4), 93.

Jiang, Z., Seyedi, S., Haque, R. U., Pongos, A. L., Vickers, K. L., Manzanares, C. M., Lah, J. J., Allan I. L., & Clifford, G. D. (2022). Automated analysis of facial emotions in subjects with cognitive impairment. *PLoS One, 17*(1), e0262527. https://doi.org/10.1371/journal.pone.0262527

Jokinen, J. P., & Silvennoinen, J. (2020). The appraisal theory of emotion in human–computer interaction. In R. Rousi, LeikasJ. , & SaariluomaP. (Eds.), *Emotions in technology design: From experience to ethics*, (pp. 27–39). Springer.

Jordan, P. (2000). *Designing pleasurable products: An introduction to the new human factors.* Taylor & Francis.

Jundanian, K. (2007). *How does the order of incorporation of design requirement solutions impacts design outcomes?* In International Association of Societies of Design Research (Ed.), *IASDR 2007 Emerging Trends in Design Research* [Press release]. Retrieved November 11, 2007, from International Association of 50 Societies of Design Research Ondisc database.

Kaufman, B. E. (2015). Integrating emotions into economic theory. In *Handbook of contemporary behavioral economics* (pp. 100–120). Routledge.

Kaygan, P., Ilhan, A. O., & Oygür, I. (2020). Change in industrial designers' jobs: The case of Turkey, 1984–2018. *The Design Journal, 23*(6), 821–841.

Keen, S. (2015). Narrative emotions. In S. Keen (Ed.), *Narrative form* (pp. 152–161). Palgrave Macmillan.

Kim, M., & Park, S. O. (2013). Group affinity based social trust model for an intelligent movie recommender system. *Multimedia tools and applications, 64*(2), 505–516.

Kimmel, S., Landwehr, E., & Heuten, W. (2024). Kinetic connections: Exploring the impact of realistic body movements on social presence in collaborative virtual reality. *Proceedings of the ACM on Human-Computer Interaction, 8*(CSCW2), 1–30.

Khan, A. N., Khan, N. A., Bodla, A. A., & Gul, S. (2020). Impact of psychopathy on employee creativity via work engagement and negative socioemotional behavior in the public health sector: Role of abusive supervision. *Personnel Review, 49*(8), 1655–1675. https://doi.org/10.1108/PR-02-2019-0072

Kouprie, M., & Visser, F. S. (2009). A framework for empathy in design: Stepping into and out of the user's life. *Journal of Engineering Design, 20*(5), 437–448.

Kozbelt, A. (2019). Evolutionary explanations for humor and creativity. In *Creativity and humor* (pp. 205–230). Academic Press.

Krathwohl, D. R., Bloom, B. S., & Masia, B. B. (1964). *Taxonomy of Educational Objectives: Handbook 2: The Affective Domain*. Longmans, Green and Co Ltd.

Kreibig, S. D., & Gendolla, G. H. (2014). Autonomic nervous system measurement of emotion in education and achievement settings. *Psychologist, 15*(2), 142–151.

Kuhn, S. (1998). The software design studio: An exploration. *IEEE Software, 15*(2), 65–71.

Kumar, J. A., Muniandy, B., & Wan Yahaya, W. A. J. (2019). Exploring the effects of emotional design and emotional intelligence in multimedia-based learning: An engineering educational perspective. *New Review of Hypermedia and Multimedia, 25*(1–2), 57–86. https://doi.org/10.1080/13614568.2019.1596169

Lage, C. A., Wolmarans, D. W., & Mograbi, D. C. (2022). An evolutionary view of self-awareness. *Behavioural Processes, 194* (2022), 104543.

Lang, P. J. (1988). What are the data of emotion? In V. Hamilton, G. H. Bower, & N. H. Frijda (Eds.), *Cognitive perspectives on emotion and motivation* (pp. 173–191). Springer.

Lankes, M., Riegler, S., Weiss, A., Mirlacher, T., Pirker, M., & Tscheligi, M. (2008). Facial expressions as game input with different emotional feedback conditions. In InakageM. & A. D. Cheok (Eds.), *Proceedings of the 2008 international conference on advances in computer entertainment technology* (pp. 253–256). https://doi.org/10.1145/1501750.1501809

Laurans, G. F. G., Desmet, P. M., & Hekkert, P. P. (2009). Assessing emotion in interaction: Some problems and a new approach. In Université de Technologie Compiègne (Ed.), *Proceedings of the 4th international conference on designing pleasurable products and interfaces* (DPPI '09). Universite de Technologie de Compiegne.

Lave, J., & Wenger, E. (1999). Legitimate peripheral participation. In R. McCormick & C. Paechter (Eds.), *Learning and knowledge* (pp. 21–35). Paul Chapman Publishing.

Lawson, B. (2006). *How designers think: The design process demystified* (4th ed.). Elsevier/Architectural Press.

Lazarus, R. S. (1984). On the primacy of cognition. *American Psychologist, 39*(2), 124–129.

Le Roux, E., De Jong, N. P., Blanc, S., Simon, C., Bessesen, D. H., & Bergouignan, A. (2022). Physiology of physical inactivity, sedentary behaviours and non-exercise activity: Insights from the space bedrest model. *The Journal of Physiology, 600*(5), 1037–1051. https://doi.org/10.1113/JP281064

Lebbon, C., Davies, S., & Shippen, J. (2009). User centered research methods in postgraduate teaching. In A. Clarke, C. McMahon, W. Ion, & P. Hogarth (Eds.), *DS 59: Proceedings of E&PDE 2009, the 11th engineering and product design education conference–Creating a better world* (pp. 558–563). The Design Society.

LeDoux, J. (2003). The emotional brain, fear, and the amygdala. *Cellular and Molecular Neurobiology, 23*(4), 727–738.

Le Heron, R., Baker, R., & McEwen, L. (2013). Co-learning: Re-linking research and teaching in geography. In *Active learning and student engagement* (pp. 11–21). Routledge.

Lemay Jr, E. P., & Dobush, S. (2015). When do personality and emotion predict destructive behavior during relationship conflict? The role of perceived commitment asymmetry. *Journal of Personality, 83*(5), 523–534.

Lewis, M. (1995). Self-conscious emotions. *American Scientist, 83*(1), 68–78. http://www.jstor.org/stable/29775364

Li, B., Hu, M., Chen, X., & Lei, Y. (2021). The moderating role of anticipated regret and product involvement on online impulsive buying behavior. *Frontiers in Psychology, 12*, 732459.

Li, J. (2022). Research status of autism spectrum disorders related to virtual reality technology in China. In S. M. Yang (Ed.), *Third international conference on artificial intelligence and electromechanical automation* (Vol. 12329, pp. 15–19). https://doi.org/10.1117/12.2646794

Li, S., & Deng, W. (2020). Deep facial expression recognition: A survey. *IEEE Transactions on Affective Computing, 13*(3), 1195–1215. https://doi.org/10.1109/TAFFC.2020.2981446

Li, W., Ren, G., Huang, Y., Su, J., Han, Y., Li, J., Chen, X., Cao, K., Chen, Q., Shou, P., Zhang, L., Yuan, Z. R., Roberts, A. I., Shi, S., Le, A. D., & Shi, Y. (2012). Mesenchymal stem cells: A double-edged sword in regulating immune responses. *Cell Death & Differentiation, 19*(9), 1505–1513.

Liang, Z., & Jin, W. (2011, September). Design for emotion classification and retrieval of video based on user experience. In *2011 International Conference on Electronics, Communications and Control (ICECC)* (pp. 802–805). IEEE.

Liem, A., & Sigurjonsson, J. B. (2011). Adapting industrial design education to future challenges of higher education. In S. J. Culley, B. J. Hicks, T. C. McAloone, T. J. Howard, & B. Ion (Eds.), *Proceedings of the 18th International Conference on Engineering Design (ICED 11), Vol. 8. Design Education* (pp. 194–204). Design Society.

Lim, Y., Giacomin, J., & Nickpour, F. (2021). What is psychosocially inclusive design? A definition with constructs. *The Design Journal, 24*(1), 5–28.

Liu, Y., & Sourina, O. (2012, July). EEG-based dominance level recognition for emotion-enabled interaction. In *2012 IEEE International Conference on Multimedia and Expo* (pp. 1039–1044). IEEE.

Longueville, B., & Gardoni, M. (2003). A survey of context modeling: approaches, theories and use for engineering design researches. In *DS 31: Proceedings of ICED 03, the 14th International Conference on Engineering Design, Stockholm* (pp. 437–438).

Lopes, J. M., Silva, L. F., Massano-Cardoso, I., & Galhardo, A. (2025). Running towards a better brand attitude: How gamification in nike run club can help? *Journal of the Knowledge Economy, 16*, 1–29.

Luckman, J. (1967). An approach to the management of design. *Journal of the Operational Research Society, 18*(4), 345–358. https://doi.org/10.1057/jors.1967.68

Lum, B. J. (1997). Student mentality: Intentionalist perspectives about the principal. *Journal of Educational Administration, 35*(3), 210–233.

Mahan, T. J. (2021). *Design for inspectability: Investigating the effect of inspectability constraints on designs and design strategies.* The Pennsylvania State University.

Marin-Alvarez, M., Rossa-Sierra, A., & Gonzalez, E. L. (2020). Emotional design and human factors design as tool for understanding efficiency in information design process at medical documents. In G. D. Bucchianico, C. S. Shin, S. Shim, S., Fukuda, G., Montagna, & CarvalhoC. , C. (Eds.), *Advances in industrial design: Proceedings of the AHFE 2020 virtual conferences on design for inclusion, affective and pleasurable design, interdisciplinary practice in industrial design, Kansei engineering, and human factors for apparel and textile engineering* (Vol. 1202, p. 157). Springer Nature.

Markussen, T. (2017). Disentangling 'the social' in social design's engagement with the public realm. *CoDesign, 13*(3), 160–174.

Martinez, M. E. (2010). *Learning and cognition: The design of the mind.* Pearson Education.

Mason, M. (2010). Sample size and saturation in PhD studies using qualitative interviews. *Forum Qualitative Sozialforschung/Forum: Qualitative Social Research, 11*(3), 1–19. http://dx.doi.org/10.17169/fqs-11.3.1428

Mastria, S., Agnoli, S., & Corazza, G. E. (2019). How does emotion influence the creativity evaluation of exogenous alternative ideas? *PLoS One, 14*(7), e0219298.

Mathews, J. M. (2010). Using a studio-based pedagogy to engage students in the design of mobile-based media. *English Teaching: Practice and Critique, 9*(1), 87–102.

Mattingly, V., & Kraiger, K. (2019). Can emotional intelligence be trained? A meta-analytical investigation. *Human Resource Management Review, 29*(2), 140–155. https://doi.org/10.1016/j.hrmr.2018.03.002

Mauss, I. B., & Robinson, M. D. (2009). Measures of emotion: A review. *Cognition & Emotion*, *23*(2), 209–237. https://doi.org/10.1080/02699930802204677

McClelland, M. M., Ponitz, C. C., Messersmith, E. E., & Tominey, S. (2010). Self-regulation: Integration of cognition and emotion. In W. F. Overton & R. M. Lerner (Eds.), *The handbook of life-span development, Vol. 1. Cognition, biology, and methods* (pp. 509–553). John Wiley & Sons.

McClenaghan, N., Kimura, A., & Stark, L. R. (2007). An evaluation of the M&S technologies smart system II for visual acuity measurement in young visually-normal adults. *Optometry and Vision Science*, *84*(3), 218–223.

McLennan, J. F. (2004). *The philosophy of sustainable design: The future of architecture.* Ecotone Publishing.

Mehrabi Boshrabadi, A., & Hosseini, M. R. (2020). Designing collaborative problem solving assessment tasks in engineering: An evaluative judgement perspective. *Assessment & Evaluation in Higher Education*, *46*(4), 1–15.

Mehrabian, A. (2017). *Nonverbal communication*. Routledge.

Mennin, D. S., Fresco, D. M., O'Toole, M. S., & Heimberg, R. G. (2018). A randomized controlled trial of emotion regulation therapy for generalized anxiety disorder with and without co-occurring depression. *Journal of Consulting and Clinical Psychology*, *86*(3), 268–281. https://doi.org/10.1037/ccp0000289

Miles, M. B., Huberman, M., & Saldana, J. (2014). Qualitative data analysis: a methods sourcebook (Edition 3.). SAGE.

Murty, P., & Purcell, T. (2007). *Designerly, reflective and insightful ways of design.* In International Association of 50 Societies of Design Research (Ed.), *International Association of 50 Societies of Design Research* [Press release]. Retrieved from International Association of 50 Societies of Design Research Ondisc database.

Naismith, L. M., & Lajoie, S. P. (2018). Motivation and emotion predict medical students' attention to computer-based feedback. *Advances in Health Sciences Education*, *23*(3), 465–485.

Naseem, K. (2018). Job stress, happiness and life satisfaction: The moderating role of emotional intelligence empirical study in telecommunication sector Pakistan. *Journal of Social Sciences and Humanity Studies*, *4*(1), 7–14.

Naqvi, N., Shiv, B., & Bechara, A. (2006). The role of emotion in decision making: A cognitive neuroscience perspective. *Association for Psychological Science*, *15*, 5.

Noffke, S., & Somekh, B. (2009). *The SAGE handbook of educational action research.* Sage Publications Ltd.

Norman, D. (2017). Design, business models, and human-technology teamwork: As automation and artificial intelligence technologies develop, we need to think less about human-machine interfaces and more about human-machine teamwork. *Research-Technology Management*, *60*(1), 26–30.

Norman, D. A. (2004). *Emotional design: Why we love (or hate) everyday things.* Basic Books.

Nunez, P. L., & Srinivasan, R. (2006). A theoretical basis for standing and traveling brain waves measured with human EEG with implications for an integrated consciousness. *Clinical Neurophysiology: Official Journal of the International Federation of Clinical Neurophysiology*, *117*(11), 2424–2435. https://doi.org/10.1016/j.clinph.2006.06.754

O'Brien, H. L., Roll, I., Kampen, A., & Davoudi, N. (2022). Rethinking (dis)engagement in human–computer interaction. *Computers in Human Behavior*, *128*(2022), 107109.

Oh, H. J., Kim, Y. S., Choi, J. K., Park, E., & Lee, S. (2011). GIS mapping of regional probabilistic groundwater potential in the area of Pohang City, Korea. *Journal of Hydrology*, *399*(3–4), 158–172.

Oliver, R. L. (1993). Cognitive, affective, and attribute bases of the satisfaction response. *Journal of Consumer Research*, *20*(3), 418–430. https://doi.org/10.1086/209358

Ortony, A., Clore, G. L., & Collins, A. (1988). *The cognitive structure of emotions*. Cambridge University Press. https://doi.org/10.1017/CBO9780511571299

Oughton, A. (2022). Building digital health tools for diabetes: How user experience research and user interface design can improve digital health adoption. In D. C. Klonoff, D. Kerr, and E. R. Weitzman (Eds.), *Diabetes digital health and telehealth* (pp. 15–27). Academic Press.

Overbeeke, C. J., & Hekkert, P. P. M. (1999). *Proceedings of the first international conference design and emotion, 3, 4 and 5 November 1999, Delft, the Netherlands*. Technische Universiteit Delft.

Owen, C. L. (1998). Design research: Building the knowledge base. *Design Studies, 19*(1), 9–20.

Oxman, R. (2001). The mind in design: A conceptual framework for cognition in design education. In C. M. Eastman, W. M. McCracken, & W. C. Newstetter (Eds.) *Design knowing and learning: Cognition in design education* (pp. 269–295). Elsevier Science.

Pahl, G., & Beitz, W. (1996). Conceptual design. In K. Wallace (Ed.), *Engineering design* (pp. 139–198). Springer.

Pampliega, A. M., & Marroquin, M. (1998). Affective education and the new Spanish educational reforms. In P. Lang, I. Menezes, Y. J. Katz (Eds.), *Affective education: A comparative perspective* (pp. 28–50). Cassell.

Park, B., Flowerday, T., & Brünken, R. (2015). Cognitive and affective effects of seductive details in multimedia learning. *Computers in Human Behavior, 44*(2015), 267–278.

Park, G. J. (2007). *Analytic methods for design practice*. Springer London.

Park, S. Y., & Baek, Y. M. (2018). Two faces of social comparison on Facebook: The interplay between social comparison orientation, emotions, and psychological well-being. *Computers in Human Behavior, 79*, 83–93.

Pawar, S., Jacques, T., Deshpande, K., Pusapati, R., & Meguerdichian, M. J. (2018). Evaluation of cognitive load and emotional states during multidisciplinary critical care simulation sessions. *BMJ Simulation & Technology Enhanced Learning, 4*(2), 87.

Pessoa, L. (2022). Emergent processes in cognitive-emotional interactions. *Dialogues in Clinical Neuroscience, 12*(4), 433–448. https://doi.org/*10.31887/DCNS.2010.12.4/lpessoa*

Picard, R. W. (2003). Affective computing: Challenges. *International Journal of Human-Computer Studies, 59*(1–2), 55–64.

Pilcher, J. J., Morris, D. M., Limyansky, S. E., & Bryant, S. A. (2022). The effect of using activity workstations on heart rate variability during complex cognitive tasks. *Journal of American College Health, 70*(3), 948–955. https://doi.org/10.1080/07448481.2020.1782919

Pine, J., & Gilmore, J. (1999). *The experience economy*. Harvard Business School Press.

Pineau, G., Jean, E., Romo, L., Villemain, F., Poupon, D., & Gorwood, P. (2022). Skin conductance while facing emotional pictures at day 7 helps predicting antidepressant response at three months in patients with a major depressive episode. *Psychiatry Research, 309*(2022), 114401. https://doi.org/10.1016/j.psychres.2022.114401

Plutchik, R. (1994). *The psychology and biology of emotion*. Harper Collins.

Qin, C., Wong-Parodi, G., & Fan, B. (2020). Different preferences for recovery options of residential fire disasters: The effect of decision role and stressed emotion. *International Journal of Disaster Risk Reduction, 43*(2020), 101383.

Rangel-Rodríguez, G. A., Badia, M., & Blanch, S. (2021). Encouraging emotional conversations in children with complex communication needs: An observational case study. *Frontiers in Psychology, 12*(2021), 2225.

Rasouli, N., Rasoolimanesh, S. M., Rahmani, A. K., Momayez, A., & Torabi, M. A. (2022). Effects of customer forgiveness on brand betrayal and brand hate in restaurant service failures: Does apology letter matter? *Journal of Hospitality Marketing & Management, 31*(3), 1–26.

Rathore, A. K., & Ilavarasan, P. V. (2020). Pre-and post-launch emotions in new product development: Insights from twitter analytics of three products. *International Journal of Information Management, 50*(2020), 111–127.

Reeve, J. (2018). *Understanding motivation and emotion.* John Wiley & Sons.

Reinholt, N., Hvenegaard, M., Christensen, A. B., Eskildsen, A., Hjorthøj, C., Poulsen, S., Arendt, M. B., Rosenberg, N. K., Gryesten, J. R., Aharoni, R. N., Alrø, A. J., Christensen, C. W., & Arnfred, S. M. (2022). Transdiagnostic versus diagnosis-specific group cognitive behavioral therapy for anxiety disorders and depression: A randomized controlled trial. *Psychotherapy and Psychosomatics, 91*(1), 36–49.

Ren, Y., Cai, Z., Li, Y., & Han, B. (2026). Impact of campus visual natural landscape structure on adolescents' emotions. *Ecological Frontiers.* https://doi.org/10.1016/j.ecofro.2026.02.004.

Rendina, H. J., López-Matos, J., Wang, K., Pachankis, J. E., & Parsons, J. T. (2019). The role of self-conscious emotions in the sexual health of gay and bisexual men: Psychometric properties and theoretical validation of the Sexual Shame and Pride Scale. *The Journal of Sex Research, 56*(4–5), 620–631.

Restrepo, J., & Christiaans, H. (2004). Problem structuring and information access in design. *Journal of Design Research, 4*(2), 218–236.

Reymen, I. M. M. J., Hammer, D. K, Kroes, P. A., Aken, J. E., van Dorst, C. H., Bax, M. F. T., & Basten, T. (2006). A domain-independent descriptive design model and its application to structured reflection on design processes. *Research in Engineering Design, 16*(4),147–173.

Rodrigues, A. P., Jorge, F. E., Pires, C. A., & António, P. (2019). The contribution of emotional intelligence and spirituality in understanding creativity and entrepreneurial intention of higher education students. *Education + Training, 61*(7/8), 870–894.

Rodrigues, I., Lopes, J. M., Borges, A., Oliveira, J., & Oliveira, M. (2021). How can Gamified applications drive engagement and brand attitude? The case of Nike run club application. *Administrative Sciences, 11*(3), 92.

Rogoff, B. E., & Lave, J. E. (1984). *Everyday cognition: Its development in social context.* Harvard University Press.

Roseman, I. J., Mattes, K., & Redlawsk, D. P. (2021). How the emotion of contempt can help explain political effects of incivility. In I. J. Roseman, K. Mattes, & D. P. Redlawsk (Eds.), *Political incivility in the parliamentary, electoral and media arena* (pp. 107–123). Routledge.

Rowland, S. (2000). *The enquiring university teacher.* Open University Press.

Russell, J. (1983). Pancultural aspects of the human conceptual organization of emotions. *Journal of Personality and Social Psychology, 45*(6), 1281–1288.

Salas, C. E., Radovic, D., & Turnbull, O. H. (2012). Inside-out. *Emotion, 12*(3), 568–578. https://doi.org/10.1037/a0025811

Salama, A. M. (2009). *Transformative pedagogy in architecture and urbanism.* Umbau-Verlag.

Salminen, M., Hamari, J., & Ravaja, N. (2021). Empathizing with the end user: Effect of empathy and emotional intelligence on ideation. *Creativity Research Journal, 33*(2), 191–201.

Sari, P., Bulantika, S. Z., Dewantari, T., & Rimonda, R. (2020). Effects of stress coping and emotion regulation on student academic stress. *Konseli: Jurnal Bimbingan Dan Konseling (E-Journal), 7*(1), 73–80.

Saxena, A., Khanna, A., & Gupta, D. (2020). Emotion recognition and detection methods: A comprehensive survey. *Journal of Artificial Intelligence and Systems*, *2*(1), 53–79.

Schachter, S., & Singer, J. (1962). Cognitive, social, and physiological determinants of emotional state. *Psychological Review*, *69*(5), 379.

Scherer, K. R. (2005). What are emotions? And how can they be measured?. *Social Science Information*, *44*(4), 695–729.

Scherer, K. R., & Ekman, P. (2014). *Approaches to emotion*. Psychology Press. https://doi.org/10.4324/9781315798806.

Schmitt, B. H. (1999). *Experiential marketing*. The Free Press.

Schön, D. A. (1987). *Educating the reflective practitioner: Toward a new design for teaching and learning in the professions*. Jossey-Bass.

Schrage, K. M., Maxwell, J. A., Impett, E. A., Keltner, D., & MacDonald, G. (2020). Effects of verbal and nonverbal communication of affection on avoidantly attached partners' emotions and message receptiveness. *Personality and Social Psychology Bulletin*. Advance online publication.

Silvia, P. J. (2005). Cognitive appraisals and interest in visual art: Exploring an appraisal theory of aesthetic emotions. *Empirical Studies of the Srts*, *23*(2), 119–133.

Singleton, A. (2025). Using walking approaches and site-specific performance to reveal layers of feeling attached to place. *GeoHumanities*, *11*(1), 36–53.

Shadle, S. E., Marker, A., & Earl, B. (2017). Faculty drivers and barriers: Laying the groundwork for undergraduate STEM education reform in academic departments. *International Journal of STEM Education*, *4*(1), 1–13. https://doi.org/10.1186/s40594-017-0062-7

Shirai, M., Soshi, T., & Suzuki, N. (2021). Knowledge of sadness: Emotion-related behavioral words differently encode loss and failure sadness. *Current Psychology*, *40*(2021), 895–909. https://doi.org/10.1007/s12144-018-0010-9

Smith, D. L., & Fernhall, B. (2022). *Advanced cardiovascular exercise physiology*. Human Kinetics.

Soleymani, M., Asghari-Esfeden, S., Fu, Y., & Pantic, M. (2015). Analysis of EEG signals and facial expressions for continuous emotion detection. *IEEE Transactions on Affective Computing*, *7*(1), 17–28.

Solmi, M., Veronese, N., Galvano, D., Favaro, A., Ostinelli, E. G., Noventa, V., Favaretto, E., Tudor, F., Finessi, M., Shin, J.I., Smith, L., Koyanagi, A., Cester, A., Bolzetta, F., Cotroneo, A., Maggi, S., Demurtas, J., De Leo, D., Trabucchi, M. (2020). Factors associated with loneliness: An umbrella review of observational studies. *Journal of Affective Disorders*, *271*(2020), 131–138.

Soto Hormazába, M. (2021). *Emotional skills for service designers in co-creation practices*. University of Lapland. http://urn.fi/URN:ISBN:978-952-337-242-9

Southward, M. W., Sauer-Zavala, S., & Cheavens, J. S. (2021). Specifying the mechanisms and targets of emotion regulation: A translational framework from affective science to psychological treatment. *Clinical Psychology: Science and Practice*, *28*(2), 168–182. https://doi.org/10.1037/cps0000003

Spendlove, D. (2007). A conceptualisation of emotion within art and design education: A creative, learning and product-orientated triadic schema. *International Journal of Art & Design Education*, *26*(2), 155–166.

Spendlove, D. (2009). *Emotional literacy*. Bloomsbury Publishing.

Streiner, D. L., & Norman, G. R. (2011). Correction for multiple testing: Is there a resolution? *Chest*, *140*(2011), 16–18. https://doi.org/10.1378/chest.11-0523

Suchman, L. (2003). *Located accountabilities in technology production*. Lancaster University.

Sweller, J., Van Merrienboer, J. J., & Paas, F. G. (1998). Cognitive architecture and instructional design. *Educational Psychology Review*, *10*(3), 251–296. https://doi.org/10.1023/a:1022193728205

Squire, K. D. (2008). Video game–based learning: An emerging paradigm for instruction. *Performance Improvement Quarterly*, *21*(2), 7–36. https://doi.org/10.1002/piq.20020.

Tashjian, S. M., Fedrigo, V., Molapour, T., Mobbs, D., & Camerer, C. F. (2022). Physiological responses to a haunted-house threat experience: Distinct tonic and phasic effects. *Psychological Science*, *33*(2), 236–248. https://doi.org/10.1177/09567976211032231

Tesch, R. (1990). *Qualitative research: Analysis types and software tools*. Falmer Press / Routledge.

Tewari, A. Theobalt, C., Goldman, D. B., Shechtman, E., Wetzstein, G., Saragih, J., Zhu, J. Y., Thies, J., Sunkavalli, K., Agrawala, M., Niessner, M., Zollhöfer, M., Fried, O., Brualla, R. M., Pandey, R. K., Fanello, S., Lombardi, S., Simon, T., & Sitzmann, V. (2020). State of the art on neural rendering. *Computer Graphics Forum*, *39*(2), 701–727.

Tian, Y. I., Kanade, T., & Cohn, J. F. (2001). Recognizing action units for facial expression analysis. *IEEE Transactions on Pattern Analysis and Machine Intelligence*, *23*(2), 97–115. https://doi.org/10.1109/34.908962

Topçu, S., & Leana-Taşcılar, M. Z. (2018). The role of motivation and self-esteem in the academic achievement of Turkish gifted students. *Gifted Education International*, *34*(1), 3–18.

Tracey, M. W., & Hutchinson, A. (2019). Empathic design: Imagining the cognitive and emotional learner experience. *Educational Technology Research and Development*, *67*(5), 1259–1272.

Tracy, J. L., & Robins, R. W. (2008). The automaticity of emotion recognition. *Emotion*, *8*(1), 81.

Trigwell, K., Prosser, M., & Waterhouse, F. (1999). Relations between teachers' approaches to teaching and students' approaches to learning. *Higher Education*, *37*(1999), 57–70. https://doi.org/10.1023/A:1003548313194

Tronstad, C., Amini, M., Bach, D. R., & Martinsen, O. G. (2022). Current trends and opportunities in the methodology of electrodermal activity measurement. *Physiological Measurement*, *43*(2), 1–16. https://doi.org/10.1088/1361-6579/ac5007.

Turner, F. (2021). *From counterculture to cyberculture*. University of Chicago Press.

Tzvetanova, S. (2007). Emotional interface methodology. In International Association of Societies of Design Research (Ed.), *IASDR 2007 emerging trends in design research* [Press release]. Retrieved November 4, 2007, from International Association of 50 Societies of Design Research Ondisc database.

Tzvetanova, S., Tang, M.-X., & Justice, L. (2007). Emotional web usability evaluation. In J. A. Jacko (Ed.), *12th International Conference, HCI International 2007, Beijing, China, July 22–27, 4553*(2007), 1039–1046.

Uwajeh, P. C., Iyendo, T. O., & Polay, M. (2019). Therapeutic gardens as a design approach for optimising the healing environment of patients with Alzheimer's disease and other dementias: A narrative review. *Explore*, *15*(5), 352–362.

Uzun, A. M., & Yıldırım, Z. (2018). Exploring the effect of using different levels of emotional design features in multimedia science learning. *Computers & Education*, *119*(2018), 112–128.

Vagnoni, E., Lourenco, S. F., & Longo, M. R. (2012). Threat modulates perception of looming visual stimuli. *Current Biology*, *22*(19), R826–R827. https://doi.org/10.1016/j.cub.2012.07.053

Van Aken, J. E. (2005). Management research as a design science: Articulating the research products of mode 2 knowledge production in management. *British Journal of Management*, *16*(1), 19–36.

Van Dooren, E., Boshuizen, E., Van Merriënboer, J. J. G., Asselbergs, T., & Van Dorst, M. (2014). Making explicit in design education: Generic elements in the design process. *International Journal of Technology and Design Education*, *24*(1), 53–71. https://doi.org/10.1007/s10798-013-9246-8

Van Gorp, T., & Adams, E. (2012). *Design for emotion*. Elsevier.

Van Merriënboer, J. J. G., & Sweller, J. (2005). Cognitive load theory and complex learning: Recent developments and future directions. *Educational Psychology Review*, *17*(2), 147–177.

Vaughan, S., Austerlitz, N., Blythman, M., Grove-White, A., Jones, B. A., Jones, C. A., Morgan, S., Orr, S., Shreeve, A.(2008). Mind the gap: Expectations, ambiguity and pedagogy within art and design higher education. In L. Drew (Ed.), *The student experience in art and design in higher education* (pp. 125–148). Jill Rogers Associates Limited, Cambridge.

Vesisenaho, M., Juntunen, M., Häkkinen, P., Pöysä-Tarhonen, J., Fagerlund, J., Miakush, I., & Parviainen, T. (2019). Virtual reality in education: Focus on the role of emotions and physiological reactivity. *Journal of Virtual Worlds Research*, *12*(1), 1–14.

Violante, M. G., Marcolin, F., Vezzetti, E., Nonis, F., & Moos, S. (2019). Emotional design and virtual reality in Product Lifecycle Management (PLM). In P. Ball, L. H. Huatuco, R. J. Howlett, & SetchiR. (Eds.), *Proceedings of the 6th International Conference on Sustainable Design and Manufacturing (KES-SDM 19).*

Wagstaff, C. R., Hanton, S., & Fletcher, D. (2013). Developing emotion abilities and regulation strategies in a sport organization: An action research intervention. *Psychology of Sport and Exercise*, *14*(4), 476–487.

Walther, J., Miller, S. E., & Sochacka, N. W. (2017). A model of empathy in engineering as a core skill, practice orientation, and professional way of being. *Journal of Engineering Education*, *106*(1), 123–148.

Wang, T., & Zhou, M. (2020). A method for product form design of integrating interactive genetic algorithm with the interval hesitation time and user satisfaction. *International Journal of Industrial Ergonomics*, *76*(2020), 102901.

Wegner, D. (2002). *The illusion of conscious will*. MIT Press.

Weilenmann, S., Schnyder, U., Parkinson, B., Corda, C., Von Kaenel, R., & Pfaltz, M. C. (2018). Emotion transfer, emotion regulation, and empathy-related processes in physician-patient interactions and their association with physician well-being: A theoretical model. *Frontiers in Psychiatry*, *9*(2018), 389.

Weiss, N. H., Dixon-Gordon, K. L., Peasant, C., & Sullivan, T. P. (2018). An examination of the role of difficulties regulating positive emotions in posttraumatic stress disorder. *Journal of Traumatic Stress*, *31*(5), 775–780. https://doi.org/10.1002/jts.22330

White, C. J., Johnson, K. A., & Mirbozorgi, B. (2023). Unique cognitive and emotional profiles of interpersonal gratitude and spiritual gratitude. *The Journal of Positive Psychology*, *127*(3), 497–517.

Williams, P. G., Johnson, K. T., Bride, D. L., Baucom, B. R. W., & Crowell, S. E. (2022). Individual differences in aesthetic engagement and proneness to aesthetic chill: Associations with awe. *Psychology of Aesthetics, Creativity, and the Arts*. Advance online publication. https://doi.org/10.1037/aca0000458

Wood W. B. (2003). Inquiry-based undergraduate teaching in the life sciences at large research universities: a perspective on the Boyer Commission Report. *Cell Biology Education*, *2*(2), 112–116. https://doi.org/10.1187/cbe.03-02-0004.

Wormald, P. W., & Rodber, M. (2008). Aligning industrial design education to emerging trends in professional practice and industry. *Journal of Design Research*, *7*(3), 294–303.

Xue, H., Desmet, P. M., & Fokkinga, S. F. (2020). Mood granularity for design: Introducing a holistic typology of 20 mood states. *International Journal of Design*, *14*(1), 1–18.

Yang, G., Chen, J., Gao, Z., Li, S., Ni, H., Angelini, E.,Wong, T., Mohiaddin, R., Nyktari, E., Wage, R., Xu, L., Zhang, Y. Du, X., Zhang, H., Firmin, D.,KeeganJ., (2020). Simultaneous left atrium anatomy and scar segmentations via deep learning in multiview information with attention. *Future Generation Computer Systems*, *107*(2020), 215–228.

Yim, B. H., & Byon, K. K. (2018). The influence of emotions on game and service satisfaction and behavioral intention in winning and losing situations: Moderating effect of identification with the team. *Sport Marketing Quarterly*, *27*(2), 93–107 .

Young, L. E. (2019). The psychology of state repression: Fear and dissent decisions in Zimbabwe. *American Political Science Review*, *113*(1), 140–155.

Zajonc, R. B. (1998). Emotions. In D. T. Gilbert, S. T. Fiske, & G. Lindzey (Eds.), *The handbook of social psychology, Vol. 1–2* (4th ed., pp. 591–632). McGraw-Hill.

Zaki, J. (2020). Integrating empathy and interpersonal emotion regulation. *Annual Review of Psychology*, *71*(2020), 517–540.

Zanganeh, M. Y., & Hariri, N. (2018). The role of emotional aspects in the information retrieval from the web. *Online Information Review*, *42*(4), 520–534.

Zhou, J. J., Phadnis, V., & Olechowski, A. (2021). Analysis of designer emotions in collaborative and traditional computer-aided design. *Journal of Mechanical Design*, *143*(2), 1–12.

Zoltowski, C. B., Oakes, W. C., & Cardella, M. E. (2012). Students' ways of experiencing human-centered design. *Journal of Engineering Education*, *101*(1), 28–59.

Zotev, V., Mayeli, A., Misaki, M., & Bodurka, J. (2020). Emotion self-regulation training in major depressive disorder using simultaneous real-time fMRI and EEG neurofeedback. *NeuroImage: Clinical*, *27*, 102331.

Index

For Product Safety Concerns and Information please contact our EU representative GPSR@taylorandfrancis.com Taylor & Francis Verlag GmbH, Kaufingerstraße 24, 80331 München, Germany

Batch number: 10476195

Printed by Printforce, the Netherlands